BUILD YOUR DREAM HOUSE FOR A SONG

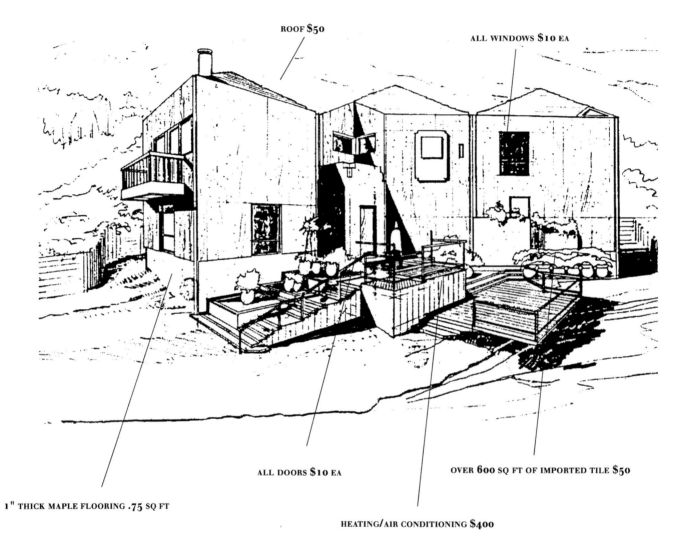

ROOF $50

ALL WINDOWS $10 EA

1" THICK MAPLE FLOORING .75 SQ FT

ALL DOORS $10 EA

OVER 600 SQ FT OF IMPORTED TILE $50

HEATING/AIR CONDITIONING $400

LAND 20 CENTS ON THE DOLLAR

BUILD YOUR DREAM HOUSE FOR A SONG

AND OWN IT FREE AND CLEAR IN FIVE YEARS OR LESS

DAVID COOK. OWNER BUIILDER.

Illustrations by GREG KLINGER.

WARBECK PUBLISHING SAN RAFAEL, CALIFORNIA

WARNING — DISCLAIMER

All advice and information in this manual is given in
good faith and in no way replaces state or local
building requirements and the need for individually
prepared plans and specifications. The author and
publisher therefore disclaim all legal liability should
any advice or information be incorrect, incomplete,
inappropriate or in any way defective. It should be
further understood that the author and publisher are
not engaged in rendering legal, accounting or other
professional services. If legal or other expert
assistance is required, the services of a competent
professional should be sought. There may be
mistakes both typographical and in content. There
fore, this book should be used only as a general guide
and not as the ultimate guide to house building. The
purpose of this manual is to educate and entertain.
The author and publisher shall have neither liability
nor responsibility to any person or entity with
respect to any loss, damage or injury caused or
alleged to be caused directly or indirectly by the
information contained in this book.

Warbeck Publishing
P.O. Box 3341
San Rafael, California 94912

Cook, David
Build Your Dream House for a Song
and own it free and clear in five years / Cook David
includes bibliographical reference and index
ISBN 1-890824-36-4

Printed in the United States of America

How does it feel
    to be without a home
like a complete unknown
    like a rolling stone?
BOB DYLAN *Like a Rolling Stone* (Song)

The book is dedicated to
Owner Builders everywhere
past present and future

CONTENTS

ACKNOWLEDGMENTS

I wish to thank my wife Marcia, and my daughters, Emily and Liz, for their help and unequivocal support during the writing of this book, and John Abbott for teaching me how to build a house.

Several people helped in the completion of this book. I am indebted to all of them: Sharon Hennessy, who did the initial edit, my daughter Emily Cook who did the second edit and Hennessy Knoop who did the final and most grueling edit, and Steve Renick for his guidance regarding book design. Without them I would still be a long way from publishing.

I am obliged to the late Robert Oliver of Oliver's Books in San Anselmo, CA, for telling me about The Bay Area Independent Publishers Association (formally M.S.P.A.) where I gained insight into the world of self publishing, and Dan Poynter who helped steer me in the right direction.

I must thank Gulimerlo Marin for designing an incredible house; Greg Klinger for all the line drawings in this book; Peck Drennan for the architectural drawings; Stefanie Marlis for advertising copy consultation; Christopher Mitchell for computer help; Ronald Starr for his help with tax defaulted land sales; Robert Bruss for his great class on California real estate law; Rob Roy, for his assistance on owner-builder schools; Sam Lawson and the staff at Christmas in April, San Francisco and especially Maureen Carew for getting me involved; Robert "Skip" Weahunt and Debra Verniest of the Owner Builder Center in Sacramento, for their assistance and in particular their help regarding construction loans; George Kiskaddon of the Builders Booksource in Berkeley CA, for his help in the preparation of the list of suggested reading; Don Zea V.P. California Forestry Association; and Fowles Auction Group with Overetts Rocklea, Qld Australia for allowing me to take photos at one of their auction sales.

INTRODUCTION

L ike most people I used to think that in order to have a house of your own you had to scrimp and save until you had enough for a down payment, and then sign up for a thirty year mortgage. That's what I did, not realizing at the time that it was really more the bank's house than my own. After a few years of making mortgage payments I decided there had to be a better way. So I made up my mind to build my own house—not just any house either, but my *dream house*— and I was determined to do it as cheaply as possible. When the dust settled and I added everything up, I discovered I'd saved so much money that I actually made over a quarter million dollars in equity.

The thirty year mortgage is a phenomenon of only the past two generations, encouraged by institutional lenders and fueled by the mortgage interest deduction. Those who are industrious enough to save up a down payment for a house, then have the pleasure of working the next thirty years to pay back the lender—a figure which ends up more than double the original price of their home. If they can't make the payments, the lender takes the house.

Say you have an average goal in the conventional housing market of a standard three-bedroom tract-built house, you probably have the where-withal to build an exceptional house. You could take your down payment and stretch the spending power of those dollars.

Your house is more than likely your single biggest investment. It may in fact be the reason you are working so hard—just to make your mortgage payments. Imagine how different your life would be if you had no mortgage payment and no rent?

You can live in a house that you own outright, and it can be *your* dream house. Thanks to modern technology, building your own house has never been easier. Nail guns are available to do every kind of nailing job, electric saws to do every kind of cutting job and a myriad of other helpful tools that were not available a generation ago.

The impetus for this book was the ease at which I, with almost no experience, built *my* dream house. It was a lot of work, but the end result was well worth it. Imagine owning your dream home free and clear!

In these pages you will learn how to save money every step of the way. However, saving money in only a couple of areas will still allow you to beat the conventional system of home ownership and have the house of your dreams free and clear in five years or less.

During the course of building my own dream house for as little as possible and the subsequent research for this book, I discovered five major areas for saving a great deal of money, and I will cover them in detail in the following chapters. They are:

How to   buy land for as little as twenty cents on the retail dollar.
buy building materials for as little as ten cents on the dollar.
design for ease of construction, thereby saving a lot of money.
live on your land comfortably and rent free while you build.
get the best help for the least money.

1 HISTORICAL PERSPECTIVE

Since earliest times, we have adapted our shelters to climate and available materials. When the first European settlers landed in North America, they built temporary dwellings using what they could find. The half-faced camp on the previous page was a typical temporary shelter. The roof was little more than poles or saplings covered with earth and thatch. Once they had established their base camps, the settlers were able to build permanent dwellings like the ones they had left back in their homelands. Settlers from Scandinavia and Germany built log houses, while those from Britain built houses in a traditional timber-frame style (fig. 1).

To start their timber-frame houses, settlers cut down trees, shaped them with an adz, and then fastened the squared timbers together with mortise and tenon joints. For the roof, they used thatch or split-wood shakes. And to cover the space between the timbers, they used wattle and daub, a primitive system of mud or clay trowelled on interwoven branches. Later, they replaced the wattle and daub with clapboards. Only the rich could afford glass for their windows as this was an expensive commodity imported from Europe. Instead, most people cut openings in the walls and covered them with boards at night.

During this time of closely-bound community, people often helped build one another's houses. Many of the earliest settlers' homes had only one room, but as their families grew, they added on. Many of the large houses we see today in New England were actually built in stages. Figure three shows how a settler's home might have expanded room by room.

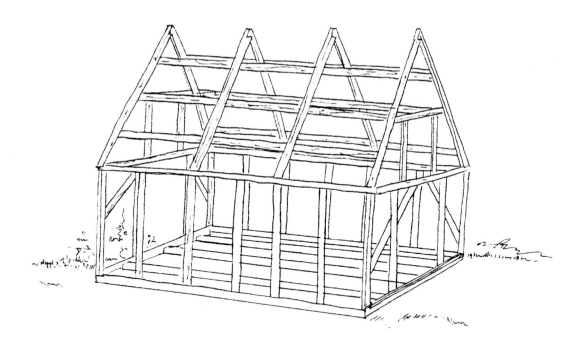

Fig. 1. Timber-frame house.

Notice that the fireplace, originally at one end of the house, found itself in the middle of the structure with the later addition—no doubt, a more practical placement for heating efficiently.

America was endowed with vast forests, and lumber mills were built up and down the Eastern Seaboard. By 1770 in New England, practically every town on a stream had a sawmill. Settlers cut trees in virgin forests, then floated them down stream to a mill, and turned them into lumber to build their houses and barns. As the settlers moved west into the Great Plains region, they found that there was no wood for building and so used sod, the only building material available. After getting established and growing crops, they built their houses from straw bales. However, wood—versatile and abundant elsewhere in the country—was to become America's principle building material.

Fig. 2. Log cabin.

fig. 3. Log cabin expansion.

12

*Fig. 4. Modern lightweight
wood-frame construction.*

LIGHTWEIGHT WOOD-FRAME CONSTRUCTION

By the middle of the 1800s, so many immigrants were flooding into the country that a fast, easy system of building houses was desperately needed. Lightweight wood-frame construction was developed to meet the demands of an ever-growing population (fig. 4).

This versatile system of framing is now used in many parts of the world and is still used almost exclusively in *tract-built* housing in North America. As long as forests are managed properly and lumber is grown and harvested for its intended purpose (I'm not suggesting we cut down old-growth forests), it seems that lightweight wood-frame construction will be a long-term, viable solution for future housing needs.

Fig. 5. Students and teachers from OUT ON BALE *in* TUCSON, ARIZONA *errecting a straw-bail building.*

HOUSING TODAY

Though the majority of houses in the U.S. are built of lightweight wood-frame construction, other methods still exist. In some places, alternative methods of building are making a comeback—a fact borne out by the number of schools across the country teaching these alternative methods (see chapter seven).

People still build timber-frame houses and log houses, and straw-bale houses are also gaining popularity as people become increasingly concerned with the environment. Straw, a frequently discarded byproduct of farming, can achieve an insulation factor of R50-R60. Compared to a standard stud or lightweight wood-frame construction, which has an insulation factor of R13-R19, a straw-bale house has significantly lower heating costs.

So it seems we've come full circle: once again using the same housing styles that our forbearers brought to this country three hundred years ago—often with increasing acceptance from building departments.

CONCLUSION

Only a small percentage of all new houses in the U.S. today are owner built; however, I believe that many more people would build their own homes if they knew how easy it is and how it could change their lives.

Imagine: not having to work nine-to-five, not having to commute, not having to pay rent on somebody else's property, and not having to feel like one of the masses shackled by a thirty-year mortgage.

Needless to say, the world that the settlers knew is very different from our own, but that doesn't mean we need to relinquish the independence and self-sufficiency that was such an integral part of their lives. Building your own home is an extraordinarily gratifying experience that lends just that—independence and self-sufficiency. Yes, it takes patience and perseverance, but the rewards are worthwhile: you live in a beautiful home, custom built to your family's needs. Your life is not ruled over by the evil demon of money-stress. And finally, yet most importantly, you have time—to spend with your family, to relax, and to pursue those activities that make you most content.

2   LAND:
FINDING AND BUYING

There are many things to consider before buying land. The land you select must be buildable; apart from that, you shouldn't let *anything* be set in stone. Everyone has a wish list, but, quite frankly, no one piece of land will have both the mountains and the plains as well as all the other qualities that will make it buildable and livable.

Make a list of priorities to help you determine which ones can be compromised. Here are eleven criteria to consider when looking for land:

1. *Cost:* terms and down payment.
2. *Employment*: Are you looking in the area of your current employment or do you plan on moving into a new area?
3. *Schools:* An important consideration if you have or plan on having children.
4. *Essential services:* Health care, police, groceries and supplies.
5. *Utilities:* Electricity, telephone, gas, water, etc..
6. *Soil:* Can you install a septic system? Or, is the land on a city sewer? Do you want to grow things? Is it contaminated and unsafe for children?
7. *Exposure:* For direct sunlight year-round, southern is best.
8. *Neighbors:* Go and meet them. Try to get an idea of what they're like—they'll be a good source for information on the area.
9. *Resale:* What is the area like? If you ever want to sell, will you make a good profit? Check both land prices and house prices in the neighborhood.
10. *Noise:* Is the site near a freeway or street with heavy traffic?
11. *Lay of the Land:* Is the lot level? How much preparation will it take before you can build?

BEST TIME TO BUY

The best time to buy land is in the fall and winter. When everything looks bleak, you can negotiate a better price. In spring and early summer, well, things are more appealing—the sun is shining, the birds are singing, people are in love, demand is high and prices are up.

ZONING

A piece of residential land is appraised by the number of houses that can be built on it, so a buildable lot is just that. The value of a parcel depends on how the land is zoned for residential purposes.

Sometimes you can split a parcel of land into two or more buildable lots. Always consider this when looking for land and check your county maps to see if other parcels close to you have been split, and if so, how recently. Soon after buying our property, we found out it could be split into two parcels, which we did. The result was a substantial profit when we sold one of the parcels.

METHODS OF BUYING

There are a number of methods for purchasing land in this country, varying from the conventional to the very unconventional.

The purchase of land from a real estate company offers the most choice whereas buying from a tax-defaulted land sale offers the least. When you are building your own dream house, I recommend that you pay a little more to buy a piece of land you really want.

REAL ESTATE AGENTS

Most people go to a real estate company in the area where they want to live, tell the agents what they are looking for, and let the agents do the work. Though it is the most expensive way to buy land, using a real estate agent has advantages. In simple terms, real estate agents provide protection for both parties. For example, if a piece of land is currently listed and you are afraid of loosing the opportunity to make an offer, you would, of course, make an offer with a real estate agent.

However, use your own agent, not the listing agent. You'll have double the protection with two real estate companies on the hook, and if you end up in a dispute with the seller, you can probably turn to your real estate agent to resolve it. If you had not enlisted the services of a real estate agent, you would probably have to hire a lawyer.

DEALING DIRECTLY WITH THE SELLER—EXPIRED LISTINGS

The commission charged by real estate agents (typically 10 percent in the U.S.) is paid for by the seller and transfers directly into the selling price. If you can wait until a listing has expired, you will probably save as much as 20 percent on the listed price.

Sellers usually list their land for more than they are willing to accept (frequently 10 percent more). So if the asking price is $50,000 and the real estate commission is 10 percent, the seller will probably settle with any offer above $45,000. If the assumption is true that the seller would have accepted an offer of $45,000 while the real estate agent was on board, the seller would have netted $40,500. Now, if after the listing has expired, you come in with an offer of $41,000 (20 percent less plus $1,000 to sweeten the deal) then the seller is getting a pretty good price—because he takes it all, and there's no middle man.

If this is your preferred strategy, avoid giving your name to the real estate agent that has the property listed as he or she may be entitled to a commission even if you purchase the property after the listing has expired. You should hire a lawyer specializing in real estate matters to handle the paperwork or use a good contract such as that in *Finding & Buying Your Place in the Country* by Les and Carol Scher (Dearborn Financial Publishing Inc).

NEWSPAPER CLASSIFIED ADS

Weekend newspapers have the highest readership and thus the most advertisements. Most sellers don't bother advertising any other time because it's not worth their while. If an ad is paid for by a real estate company, it should be indicated.

Everybody hates house-agents because they have everybody at a disadvantage. All other callings have a certain amount of give and take; the house agent simply takes.

H. G. WELLS

SIGN ON PROPERTY

The most difficult aspect of a sign on the property is assessing fair market value. Check at the county assessor's and recorder's offices to see what similar parcels have sold for in the past year or two. Also, talk to neighbors.

WORD OF MOUTH

Many people have land for sale but don't advertise the fact. This is especially true the further you get from towns and cities. It's important to let people know that you're a buyer so mention this at shops, bars, clubs, churches, and anywhere else word gets out.

STUMBLING UPON THAT DREAM SITE

Happy the man whose wish and care a few paternal acres bound, content to breathe his native air in his own ground.
ALEXANDER POPE

Perhaps, while wandering down a country road, all of the sudden, there it is—the spot for your dream house! But how do you find out who owns the land? And if they want to sell it?

If you see some land that you would like to buy and don't know whom it belongs to, go to the county recorder's office and look it up. It's helpful if you can provide a street address or some other description, but it's also possible to look at the county maps to identify the property.

Ownership and tax information on a piece of land is public record. At county offices, for example, you can see if the property taxes have been paid as well as if there are any other recorded liens and assessments. Easements are recorded too, so you will know ahead of time if the utility company has a right of way or if a neighbor has the right to gain access to his or her adjacent property.

With just a person's name you can see what they own, what liens are recorded against their property, if they've ever filed bankruptcy, and if there are any judgements recorded against their property. This sort of information can be *very* helpful in negotiations.

NOTICE OF PUBLIC AUCTION ON OCTOBER 26, 1995 OF TAX DEFAULTED PROPERTY FOR DELINQUENT TAXES

On July 18, 1995, I Frank D. Hodges, Tuolumne County Tax Collector, was directed to conduct a public auction sale by the Board of Supervisors of Tuolumne County, California. The tax defaulted properties listed below are subject to the tax collector's power of sale and an authorization to sell, dated Sept. 22, 1995 has been received from the State Controller. This is public notice that I will sell the properties at 10:00 a.m. on Thursday, October 26, 1995, in the Board of Supervisors at #2 S. Green St., Sonora, California 95370. This sale will be conducted as a public auction to the highest bidder for cash in lawful money of the United States or negotiable paper, for not less than the minimum bid as shown on this notice. Properties which are redeemed in full by October 25. 1995 will not be sold. The right of redemption will cease at that time and properties not redeemed will be sold. If the properties are sold parties of interest, as defined in California revenue and taxation code section 4675 have a right to file a claim with the county for excess of 150.00 plus proceeds from the sale. Excess proceeds are the amount of highest bid in excess of the liens and costs of the sale which are paid from the sale price. Notice will be given to parties of interest, pursuant to law, if excess proceeds result from the sale. Potential bidders should contact the tax collector at #2 S. Green St., Sonora, Ca. 95370 or call 533-5544 for more information regarding the public auction.

PARCEL NUMBERING SYSTEM EXPLANATION

The Assessor's parcel number (APN) when used to describe property in this list, refers to the assessor's map book, the map page, the block on the map, (if applicable) and the individual parcel on the map page or in the block. The assessor's maps and further explanation of the parcel numbering system are available in the assessor's office.

The properties that are the subject of this notice are situated in Tuolumne County, California and are described as follows:

Item No.	APN	Last Assessee	Min. Bid
1.	037-127-05-0	Alpheus Ponce, 1/2- Barbara G. Ponce 1/2	$8,630.00
2.	038-300-15-0	Mary N. Brown Life Estate	18,500.00
3.	038-310-39-0	Tremon E. Roberts	22,150.00
4.	062-131-26-0	Clark and Blanche Williams	2,000.00
5.	076-220-07-0	Daniel Estrada %TCE	2,250.00
6.	082-213-10-0	Frank and Shelly L. Domnick, Jr.	1,500.00
7.	083-112-05-0	Gordon D. & Mary Otter	8,100.00
8.	084-090-30-0	Art & Nadine Rasor	1,500.00
9.	087-060-04-0	John L. & Eileen R. Bliss	800.00
10.	090-310-01-0	David Scott Hallman TR.	2,500.00
11.	092-080-08-0	Steven J. Barretta 2/6 Jacquline Barretta 1/6 et al	1,100.00
12.	093-080-09-0	H.L. Behlman Jr.	1,700.00
13.	093-080-11-0	H.L. Behlman	1,800.00
14.	093-080-12-0	H.L. Behlman	1,800.00

Publication dates: Oct. 3, 10, 17, 1995. The Union Democrat, Sonora, Ca.

Fig. 6. This is a copy of the actual notice that appeared in the local newspaper. It was faxed to me by the tax collector's office. I deleted all items that had been redeemed, so only those lots still for sale remained on my list.

TAX DEFAULTED LAND SALES: BUYING LAND
FOR AS LITTLE AS TWENTY CENTS ON THE DOLLAR

A tax-defaulted land sale is a forced sale of private property by a county to recover unpaid property taxes. Most counties in the United States hold tax-defaulted land sales. However, they don't usually auction the land directly but, instead, auction off tax certificates. The certificate may be redeemed by the property owner during a set right-of-redemption period, lasting from six months to four years. If the certificate is not redeemed in that period, the holder of the tax certificate can foreclose on the property. If it is redeemed, the holder is paid interest on his or her money—sometimes as high as 12 to 16 percent.

In California and about a dozen other states these properties are sold outright at public auction. There is no redemption after the auction. In Florida, tax certificates are issued; however, the properties are also sold at public auction to the highest bidder after the two-year right-of-redemption has expired.

Counties don't like to own these properties because they are not producing taxes. In some states, you may find counties that have such properties on file and will sell them to you for the amount of back taxes owed. There are over three thousand counties in the United States, and unfortunately there are no hard rules. I understand that, in many states, the individual counties make their own rules, so it's important to do a little research. You can write to tax collectors, or telephone them, or better yet drop in on them for a visit.

In California and other tax deed states, this is how it works: the successful bidder gets the property free and clear and is issued a tax deed, which is analogous to a grant deed. Properties that are over five years delinquent in property taxes are sold at public auction to the highest bidder. This redemption period varies from state to state; however, it doesn't really matter since the bidder is only interested when the property comes up for sale. After an auction, the successful bidder pays for the property, and a tax deed is mailed to the new owner within a couple weeks. That's it. No further redemption period. The old owner ran out of luck when the redemption period ran out at 5 P.M. the day before the auction.

Except for an I.R.S. lien, which is federal, a state lien is senior to all other liens. Buying property at a tax-defaulted sale wipes out all other liens. By the way, I.R.S. liens are rare, but if the I.R.S. does have a lien, it has 120 days to notify the new owner. In any case, if this were to happen, the money paid by the new owner would be returned. The county usually knows if there's an I.R.S. lien but has no obligation to disclose the information. Nine times out of ten, if someone asks, the county will tell.

PREPARING FOR A TAX-DEFAULTED LAND SALE

I recently bought two parcels at a tax-defaulted land sale in Northern California. According to a real estate agent, one is worth $15,000; I paid $4,500. The second is worth around $10,000; I paid $2,000. This is how I went about acquiring these two buildable lots for pennies on the retail dollar.

First, I phoned several tax collectors' offices in different counties to find out if they were planning an auction of tax-defaulted properties in the near future. When I found a county that was going to hold an auction, I asked to have a list of the properties sent to me. Two days before the auction, I began doing my research. I wouldn't have wanted to start any sooner because most properties are redeemed before they go to auction. By waiting until the last two or three days, I didn't waste any time.

Upon arriving at the county offices, I marched directly over to the tax collector's office to get the most recent list of unredeemed properties. Next, I strode down to the assessor's office and obtained copies of the assessor's parcel maps or plat maps, as they're often called (see fig. 7). The woman at the counter was very friendly and offered me directions to each of the remaining three properties, giving me a choice between the most direct and the scenic route.

There were still a few more things to find out before I left the county offices. I went to the planning department to see what the zoning was on each of the parcels. They were all zoned for single family homes, which was good.

I then went across the hall to check with the environmental health department to see whether any of the properties were on a sewer line. They advised me to check with the sewer district. Then I asked if there were any records of septic system research from any previous owner. In other words, I wanted to know if there was no sewer, whether the soil would be suitable for a septic system. Unfortunately, no records were available. Next, I asked them whether the county allowed alternative septic systems for problem soils as well as if there were any areas in the

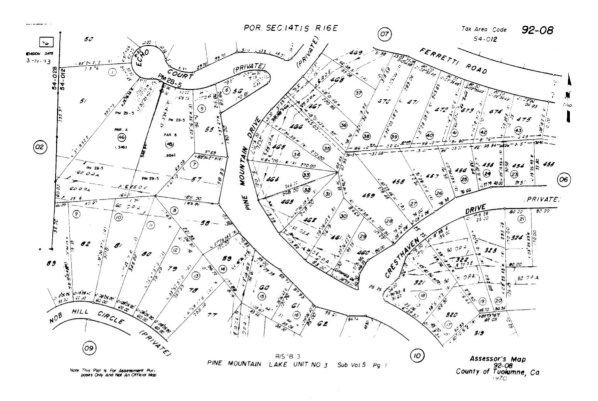

Fig. 7. I purchased lot 58 for $2,000—it's beautiful!

county that had a history of problem soils. (For if there was no sewer and if the soil was unsuitable for any kind of septic system, the land would be worthless, and I certainly wouldn't waste any more time on it.) There were no problems in the areas where I was looking. Sometimes you can't be certain about a piece of land until you actually dig a hole with a backhoe. But I'm always reassured when I see newer homes on adjacent lots. The last thing I did at the county offices was obtain a list of setbacks for the septic system from property lines, creeks, etc.

Next, I went to the building department to find out if mobile homes were permitted on these properties. (I would need a place to stay while I built.) On two of them, it was allowed; on the third, which turned out to be in a private, gated subdivision, it was not.

Later after lunch and a pleasant twenty minute drive (the scenic route), I arrived at the first property. It was in an established area with houses on both sides—just what I was hoping for. The property was set down about fifteen feet from the street, so I walked the half-acre property. The land was fairly level, with a slight slope from front to back. The soil looked sandy, which is usually a good sign for a septic system. All other essential utilities were already present. This was definitely a buildable lot.

The next lot was about a mile up the hill and had a breathtaking view. Unfortunately, the property was on the downhill side of the street, and one side of it was quite steep. But the biggest negative was that the minimum bid was $8,100. I decided to give it a miss.

The third lot was in a gated community. When I arrived at the gate, the security person didn't want to let me in without a pass, but after some fast talking, she finally let me in. This lot was also on a bit of a slope, but like the first, it was fairly level. It was set about twenty feet down from the street. I found utilities available as well, including electricity, telephone, and water. I also noticed what looked like a sewer manhole cover. Leaving this third lot, I spied a sewage treatment plant about a half mile down the road. The gate was open. Though it was after 6 P.M., I drove on in to see what I might find out.

A lone engineer was working in the office. Apologizing for bothering him after hours, I asked him if the evidence I had found in front of my subject property was, in fact, a sewer line. I showed him the assessor's plat map. He asked me to follow him upstairs where he produced the sewer district map for the same area. As it turned out, only part of the community was on the sewer line, but I was lucky. The sewer ran right in front of the lot. The engineer estimated the value of the property, "If you get it for under $10,000, you're getting a good deal."

The next morning, I still had a couple of things left to check. I wanted to know how much the monthly fees at the gated community cost. The security lady had given me a copy of the community newspaper, containing an administrative phone directory. I phoned the membership director, and found out that the dues were $47 per month—not bad, considering there was a lake, golf course, tennis courts, and other recreational facilities. The second thing I wanted to know was whether I would be required to submit house plans for design review. The answer was "yes"; the fee was $1,200. Still, the minimum bid was $1,100, and so I decided that I would pay up to $3,000. All I had left to do was wait until 10 A.M. for the auction to begin.

At 9:40 A.M. I went to the board-of-supervisors' room as instructed by the tax collector's office. Nobody was there. I thought I must be in the wrong place, because the last tax auction I attended had been very crowded. I went to the nearest office to enquire, but they knew nothing about it. I decided to go see my friends at the tax collector's office.

On my way, I met an older man—sixty-ish—and asked him if he was there for the auction. He said, "They won't be up until five minutes before ten o'clock." He told me that he lived next door to the first property I'd previewed. He had been outbid last year by a lady who apparently didn't have enough money in her bank account to cover her winning bid of $8,100. (This seemed odd since most counties require a cashier's check.) He continued, telling me that he needed the property to gain access to his own. I sensed a bit of fabrication here, and the old guy lost credibility in my book. I wasn't going to be discouraged from bidding.

Another man arrived. The old man approached him. I distanced myself, keeping within earshot. This second fellow was getting the same once over but with an added twist. The old gentleman told the new fellow that he had first right to bid on the property. The new man appeared to be taking the bait—hook, line, and sinker. A few more people arrived, and the tax collector appeared. We followed him into the supervisors' chambers.

THE AUCTION

The tax collector explained the bidding procedure: the properties would be sold to the highest bidder; no property would be sold for less than the minimum bid, which is usually the amount of owed back taxes plus some nominal fees. Sometimes assessments, such as road improvements, are also included. (If the minimum bid seems overpriced, ask the tax collector's office why.) The bidding would be in not less than one-hundred-dollar increments.

"Okay lets begin with the first lot. Do I hear a bid?"

"$1,500," said the elderly man.

This, of course, was the minimum bid. I was sitting behind the old gentleman. He couldn't see me, unless he turned around. I held up two fingers.

"We have $2,000," said the tax collector.

"$2,100," said the elderly man.

I held up three fingers.

"We have $3,000," said the tax collector.

"$3,100!" Shouted the elderly man.

I held up four fingers.

Buy land. They aren't making it anymore.
MARK TWAIN

"We have $4,000."

Now the elderly man said, "I claim ingress and egress."

"Do you hear what he's saying?" The tax collector asked me.

"I hear what he's saying, but I don't really know what he means by it," I admitted.

"Do you still want to buy the property?"

"Yes," I said.

"Okay," said the elderly man, "I'll give it one more try: $4,100."

"$4,500," I said in a cool tone, knowing it would be the winning bid.

The elderly man got up and stormed out, mumbling incoherently.

"Sold!" Said the tax collector.

I gave my name to the tax collector's assistant, and that was that.

My new neighbor had his driveway on one side of my land and he had the right to "enter and exit." Fortunately, that didn't affect the property's value. It's still a buildable lot, and I have the legal right to use the driveway.

The next lot up at the auction was the property with the breathtaking view: minimum bid $8,100 (this included a road assessment).

"Do I have a bid?"

The tax collector peered around the room. Silence.

"No bid?" He asked, "Okay, we won't sell that property today."

The final lot was the parcel in the gated community, Pine Mountain Lake. Minimum bid was $1,100. I put up my hand.

"I have $1,100," the tax collector said.

"$1,200," said another man.

I held up two fingers.

"$2,000," said the tax collector.

He waited; I waited.

"No more bids? Sold to Mr. Cook."

After the auction, I went down to the tax collector's office and paid my bill. I was given a receipt. The assistant told me that I would receive my deeds in about two weeks.

I now owned two parcels for a few dimes on the retail dollar, and it wasn't even half past ten in the morning. I strolled across the street and celebrated with a double mocha.

OFFICE OF COUNTY TAX COLLECTOR

STATE OF CALIFORNIA, COUNTY OF _____

RECEIPT FOR PROPERTY PURCHASED AT TAX SALE

Sale No. _____, Year 19 _95._   Parcel No. _92 0800_.

Receipt is hereby acknowledged of the sum of _Two Thousand + 14/00_
($), in ☐cash, ☐negotiable paper subject to collection, from

Total 2019.20

as purchase price for the following described parcel of real property which
had been deeded to the State of California for delinquent property taxes and
sold at public auction this date in accordance with Chapter 7, Park 6, Division 1, Revenue and Taxation Code:

DESCRIPTION:

_92-080-08_

After recordation, the deed will be mailed to you by the County Recorder.

_____, Tax Collector

Date _10/26/95_   _____
                          Deputy

Minimum Bid $ _1,100 00_   Recorder's Transfer Tax Collected
                           (not included in Purchase Price)
                           $ _19.20_

Please issue deed in favor of _____

and mail to _____

at _____

Receipt is acknowledged of a copy of this document which is correct in
all respects.

                                    Signed _____

Fig. 8. This is the tax sale receipt for the $2,000 lot.

CORDING REQUESTED BY:

ank D. ▮▮▮▮▮▮
eas./Tax Collector
2 S. Green St.
▮▮▮▮, Ca. ▮▮▮▮▮

lail To

vid N. Cook
▮▮▮▮▮▮▮▮▮▮▮▮▮
▮▮▮▮▮▮▮▮▮▮▮▮▮

DOCUMENT ▮   BOOK   PAGE

015128   1361 0161
▮▮▮▮▮ COUNTY OFFICIAL RECORDS

RECORDED AT THE REQUEST OF
▮▮▮▮▮▮ COUNTY TAX COLLECTOR

OCT 30, 1995   3:05:14 PM
▮▮▮▮▮▮, RECORDER
OF PAGES: 1
FEE REC'D : $19.20

Doc. Trans. Tax — computed on full value of property conveyed $ ___2.20___

Johnnie Castle
Signature of Declarant

TAX DEED TO PURCHASER OF TAX-DEFAULTED PROPERTY

On which the legally levied taxes were a lien for Fiscal Year ___1986-87___

and for nonpayment were duly declared to be in default.   ___35094___
Default Number

This deed, between the Tax Collector of ___▮▮▮▮▮▮▮___

County ("SELLER") and __David Neal Cook- Marcia Anne__

__Cook, as Community Property__ ("PURCHASER").

conveys to the PURCHASER the real property described herein which the

SELLER sold to the PURCHASER ___at Public Auction___ on ___October 26, 1995___

pursuant to a statutory power of sale in accordance with the provisions of

Division 1, Part 6, Chapter _7_, Revenue and Taxation Code, for the sum of $ ___2,000.00___.

_no_ taxing agency objected to the sale.

In accordance with law, the SELLER hereby grants to the PURCHASER that
real property situated in said county, State of California, last assessed to

▮▮▮▮▮▮▮▮▮▮▮▮▮▮▮ , described as follows: ___092-080-08-0___.
Assessors Parcel Number

ALL THAT REAL PROPERTY SITUATED IN THE UNINCORPORATED AREA OF THE COUNTY
OF TUOLUMNE, STATE OF CALIFORNIA DESCRIBED AS FOLLOWS:

Lot 58 as shown on map entitled "Pine Mountain Lake Unit No. 3"
recorded July 9, 1969 in book 5 of maps at page 1, Tuolumne
County, State of California.

STATE OF CALIFORNIA
} ss.
___▮▮▮▮▮___ COUNTY

EXECUTED ON

___October 30, 1995___   *Frank D. Hodge*
Tax Collector

On __10/30/95, Frank D. ▮▮▮▮▮, personally___, known
to me to be both the Tax Collector of said County and the person who executed this

Fig. 9. This is a copy of the actual tax deed.

TAX DEED STATES

Only a handful of states sell properties outright at tax sales, a simple phone call to the county tax collector's office will get you this information, laws sometimes change too. For example, California recently passed legislation to allow individual counties to issue tax certificates. At the present time all California counties are still issuing tax deeds at their tax-defaulted property auction sales.

TAX CERTIFICATE STATES

Generally, states that charge high interest rates on deliquent taxes are frequented by investors looking for a high return on their money. As interest accumulates, the investor stands to make a good profit when and if deliquent taxes are paid and subject property is reclaimed.

In states that charge low interest rates on tax certificates, the tax certificates are not so desirable. Few bid at these tax certificate auctions, and the county is stuck with these properties (which aren't earning any taxes).

In this second scenario, you'll find more opportunities to obtain tax-defaulted properties. If you live in a tax certificate state, check your county to see if they have any properties available. Usually, there's a fairly simple legal procedure required to gain legal title to these properties. You may need the services of a lawyer.

PUBLIC UTILITIES

Public utilities sometimes sell their surplus land. It pays to read public notices in your local newspaper. I recently saw a notice in our local newspaper from the water department, disposing of unwanted lots at public auction. The minimum bids were very low—a few hundred dollars up to $5,000.

I was interested in one lot and paid a sanitary engineer to go out and evaluate the ground. Unfortunately, the soil was unsuitable for a septic system, and thus the lot was unbuildable. If I'd purchased it without checking the soil, I might have paid thousands of dollars for a picnic spot.

NOTICE OF PUBLIC SALE

CALL FOR BIDS

In accordance with and pursuant to Government Code Sections 50569, 50570, and 54221 through 54238.6, Marin Municipal Water District (MMWD) Code Section 2.90.090 and MMWD Board Policy No. 4, the MMWD will receive bids until 6:00 p.m. on February 7, 1996, and following additional oral bids, will sell to the highest bidders by Quitclaim Deed at the Board of Directors Meeting on Wednesday, February 7, 1996, at 7:30 p.m., the following surplus parcels of land in Marin County.

APN	Approx. Sq.Ft. Area	Minimum Bid, $
Fairfax		
001-051-17	900	500
001-053-23	1,800	500
Mill Valley		
029-231-13	4,200	750
Corte Madera		
038-011-20	1,600	500
Tiburon		
038-341-04	2,500	1,000
Sausalito		
064-343-07	1,742 (5' x 348')	500
Ross		
072-021-09	300	500
Lucas Valley		
164-363-04	900	500
164-363-07	4,000	750
Lagunitas		
170-111-03	19,000	5,000
Woodacre		
172-360-17	5,000	1,000

1. All bids shall be submitted on a Bid Form in sealed envelopes obtainable from the Marin Municipal Water District and shall be accompanied by a $100 deposit. The deposit shall be in the form of a cashier's check or certified check and made payable to "Marin Municipal Water District". Deposits of unsuccessful bidders will be returned. Sealed bid envelopes will be received at the District's office, ▄▄▄▄▄▄▄▄ ▄▄▄▄▄▄▄▄▄▄▄▄▄▄▄▄▄▄▄ until 6:00 p.m. February 7, 1996.

Fig. 10. This is a Public notice of intended sale of surplus parcels belonging to the water department.

CONCLUSION

You don't have to pay retail for land! If you're careful and do your
homework, there are good unconventional strategies for buying land.
Make sure that you have a buildable lot by evaluating the soil for a
septic system and checking zoning regulations at the county offices. If
you don't feel comfortable buying land unconventionally, then buy from
a real estate agent. After all, I still have four other major money-saving
ideas for you!

3   FINANCING

During the first two hundred years of this country's history, there was little need for institutional home loans. The settlers built their homes by themselves or with the help of their communities. Most had small holdings or farms, which passed from generation to generation. If funds were needed to purchase land or for seed money, they were usually borrowed from family or friends.

Primary mortgage lending is based on the inflow of savings to institutions that would lend funds. In those early days, there was little need for secured savings, thus there was little impetus for institutionalized lending.

After the Civil War, the country industrialized and the economy grew as it never had before. Urban centers swelled with immigrants in the late 1800s, and people began to deposit their savings in banks for safe keeping. As the country expanded rapidly westward, money was needed to develop vast areas of farming land. To meet the demand for funds, a regular mortgage business sprang up in the Midwest. These mortgage companies primarily made farm loans with a loan-to-value ratio of 40 percent. As the population grew, so did the need for home mortgage money. No longer was it true that everyone was willing or able to build their own homes. In the early 1900s, the typical mortgage made on a single family residence had a loan-to-value ratio of 50 percent or less.

Today, most home mortgages have loan-to-value ratios of 80 to 90 percent with a term of thirty years. According to the National Mortgage Banker's Association, there are approximately sixty-five million home owners in the United States today—forty-two million of them have first mortgages and another eight million have second mortgages.

You don't have to be included in those statistics! If you build your own home, you might not have to borrow a cent. The best way to build your

Fig. II. This owner builder didn't use conventional financing, and the large bag of money represents sweat equity.

own house is to pay as you go. The more work you do without borrowing money; the more your dream house will pay you back in equity.

But if you can't do this, there are many options for raising funds besides borrowing from an institutional lender. In the following examples, we'll explore some financing options. If you've had difficulty borrowing conventional money, be aware of unconventional options for financing your self-built home.

CONVENTIONAL HOME LOANS

The conventional method of home ownership requires you to save up a down payment of anywhere from 10 to 20 percent or more, depending on your ability to qualify for a loan. Qualifying means that a percentage of your gross income, anywhere from 25 to 33 percent, should be available for housing expenses. You'll need to save money for closing costs as well.

This first example assumes you've made a successful offer of $150,000 on a house found by a real estate agent. In some parts of the country, the typical price will be lower, while in other areas (such as parts of California), the price will be higher. If you've saved a down payment of 20 percent (i.e. $30,000) you must still pay closing costs, adding around 6 percent to the purchase price. Closing costs include loan origination fees, credit-report fees, appraisal fees, recording and transfer fees, prorated property taxes, title search and insurance, and various other charges. Sometimes points (each point is 1 percent of the loan amount) are charged as a commission paid to a mortgage-loan broker for providing the lending institution with a new loan. Points may also be a way to keep the interest rate low by increasing the up front fees or may be charged as a penalty to a borrower with less than perfect credit (for the additional risk associated with the loan).

After thirty years of payments, this $150,000 house will end up costing $355,987. This end price includes the above mentioned closing costs and down payment, plus property taxes, which will probably be higher than those on the house you build yourself. For example, in California the property tax is about 1 percent of the purchase price. In this example, this amounts to $1,500 a year. The cost of your self-built house will be much lower than the one you buy, thus your taxes should be lower too. The taxes on the house I built were much lower than the actual value of the house.

Who borrows to build builds to sell.
CHINESE PROVERB

purchase price		150,000
down payment	20%	30,000
closing costs	6%	9,000
		--------------
loan amount		120,000
payment at 8% for 30 years = 880.52 per month		
total (not including property taxes, insurance, or maintenance)		355,987

CONSTRUCTION LOANS

Qualifying for a construction loan is similar to qualifying for a home loan—good credit, verification of employment, and sufficient income are all required. (A statement of assets and liabilities may also be requested.) In addition, you must provide the lending institution with an itemized construction cost breakdown. Expenses in this list would include: materials, labor, architect's fees, building-permit fees, surveying, insurance, clean-up, plus an amount for contingencies (things you've forgotten to include in your breakdown, but will be paid from the construction loan). Like most loans, construction loans are expensive. And they have a short term, typically less than a year. Extensions are available, as long as you don't mind paying a little more.

There are construction loan programs for owner builders. One such program, brokered by the Owner-Builder Center in Sacramento, California recognizes classroom experience to help you qualify for a construction loan. Other lenders in the area recognize this certification as well. Owner-builder schools around the country may be able to help you find construction financing.

Construction loans are paid out over several payments after various stages of construction have been completed—backwards as that may be. To build, you must convince your suppliers to wait for their money. When your house is finished, you must pay off the construction loan. This can be done with a regular thirty-year loan, not unlike a conventional home loan. Following is a construction disbursement plan supplied by the Owner-Builder Center in Sacramento:

Pre-Construction Advance. To cover approved soft costs (e.g. engineering, permits, impact fees, hookups, fire district, septic system and construction insurance). Any balance to be used as working capital in project.

1st Advance 10%. When foundation is completed; ground plumbing in place; if a slab, concrete slab poured; and C.L.T.A. 102.5 or 102.7 foundation endorsement received.

2nd Advance 10%. When subfloor is completed along with underfloor mechanics, if crawl space.

3rd Advance 15%. When building is rough framed, and roof stacked. No windows, doors, sheathing or siding.

Neither a borrower, nor a lender be; for loan oft loses both itself and friend, and borrowing dulls the edge of husbandry.

SHAKESPEARE

4th Advance 15%. When building is completely framed (including roof joists), rough plumbing complete; windows installed, exterior wrapped for stucco or exterior wood siding installed.

5th Advance 15%. When building is completed, rough H.V.A.C. (heating ventilation and air conditioning) and rough electrical completed, insulation installed, drywall hung (but not taped or textured), sheet metal completed, stucco (scratch and brown coats) completed, and all exterior doors and miscellaneous trim installed.

6th Advance 15%. When drywall is taped and textured, exterior painted or color coat completed, finish carpentry completed, interior painting completed, cabinets installed, and counter tops installed.

7th Advance 10%. When plumbing is finished, electrical is finished, H.V.A.C. is completed, all flooring installed, all appliances are in place, and building is ready for occupancy.

8th Advance 10%. The balance remaining in said account after recordation of a Notice of Completion and upon receipt of appropriate title endorsements provided there are no unpaid claims on file with Lender against the Building Loan Account. Alta rewrite received from title company.

Remember, this is an interim loan, and its purpose is to pay suppliers, subcontractors, permits and other fees. This type of loan is customarily paid off with a conventional mortgage unless the house is to be sold.

SELLER FINANCING OF LAND

Land is often financed by the seller. In this example, let's assume that you bought your land for $30,000, putting 20 percent down. And let's say you lived on your land while you're building and used your savings to purchase a mobile home (you'll make your money back because you won't be paying rent). Perhaps, you've already collected some building materials and lined up skilled help. Also let's suppose, you build in your spare time and maintain the flow of your monthly income. If you financed the land for five years at 8 percent, you'd have the following figures:

purchase price	30,000
down payment 20%	6,000
closing costs, estimate	500
-------------	-------------
loan amount	24,000

payments at 8%   for   5 years = $486.64 per month

total (including down payment and closing costs)   35,698.40

You can see that using a small short-term loan keeps both the amount of interest paid and total payments comparatively low. In fact, if you had saved up $39,000 to buy the $150,000 house in our first conventional scenario, but decided to build instead, you could purchase a used mobile home for $4,000 and still have $28,500 left over for materials and labor. Note also that if you had bought that ticky-tacky $150,000 house, you would have paid $394 more per month. These additional funds could be applied to materials and labor, thereby getting your dream house built in an even more timely fashion.

PAY AS YOU GO: NO FINANCING

This scenario assumes that you've read this book and learned about tax-defaulted land sales or have found another cheap way of buying land. It also supposes that you have decided to build your own house and don't mind spending your spare time working on your new house for the next year or two.

Purchase price,   Land   5,000   paid cash

no payments, your lot is paid in full.

Since you may have already started collecting building materials, consider erecting a shed or bringing over a container for dry storage. Design your house around the materials you've collected. Buy a mobile home or build another temporary dwelling to live in while you build your new home. Consider hiring a competent helper and then build, build, build!

You can see how far ahead you are immediately. If you do most of the work yourself and manage to get some good deals on building materials, then you'll even have some money left over from that initial $39,000 down payment. You'll own your own house free and clear and never have to pay a single mortgage payment—now that's independence!

HOW MUCH TO GET STARTED

You will need some money for the land, a temporary dwelling, and a building permit. The building permit and other related fees can easily run a few thousand dollars, or several times that depending on where you decide to build.

If you plan on building in a county that allows the permanent placement of a mobile home and don't have much money, buy the land and a mobile home first. Then build after you've saved on rent for a year or two.

MORTGAGE LOAN BROKERS

Mortgage loan brokers can help you find money—even when you've given up hope of qualifying for institutional financing. They can also find conventional financing for borrowers with less than perfect credit. They earn a commission, which is usually in the form of points for their expertise in these matters.

Mortgage loan brokers can be a valuable source of unconventional funds for partly-completed homes, although you may qualify for a conventional thirty-year mortgage if you have completed the "shell" of your home. If you build until you run out of money and can't find conventional financing, consider talking to a private mortgage loan broker. Mortgage loan brokers often have private parties willing to invest money. These investors aren't as concerned with your credit report as with the equity in your property. If they don't get their money, they foreclose, selling your property to recover their investment.

Mortgage loan brokers also make expensive, short-term loans—six months to two years. The procedure is a bit easier than other loans. The mortgage loan broker will probably come out to see the property and do the paperwork. He'll send out an appraiser to evaluate your property, and if the appraisal indicates sufficient equity, the broker will recommend your loan to his clients. Typically, 70 percent of equity is the maximum you can expect to borrow.

I would definitely try to avoid this kind of loan. It almost certainly includes points and higher interest rates than any other home loan. But if you're really in a tight spot, you may want to look into such a loan. You can usually get your money in a couple of weeks. However, when construction is completed, consider a regular home mortgage.

CONCLUSION

Most institutional lenders expect borrowers to have nearly perfect credit. Credit helps if you want to take the conventional path. But more often than not credit is a gilded cage: quick and easy money now for much too much interest later, later, later.

Indeed, the mortgage interest deduction is a tempting incentive to fall in line with the masses and sign over middle age to a thirty-year mortgage. But, remember, you'll have to earn enough money every month to make that mortgage payment just so Uncle Sam can't take away all the money that you're earning to make your mortgage. . . it's the proverbial vicious circle.

Traditional financing may not be necessary for building the house of your dreams. Take advantage of seller financing for the land; live on your property, and use the money saved from rent to pay for building materials and labor, and do a lot of the work yourself (it's not that hard—buy a nail gun).

I know that you can build your own home without financing—or at the most, short-term seller financing on the land. I know because I've built my own home this way. And in the next chapters I'm going to tell you how to do things a little differently for a great deal less.

Credit is comforting, and we may all need to borrow money from time to time. By owning our home free and clear, we can tap that equity and always get the safest and cheapest money out there, avoiding long-term loans, high interest rates, and points.

4 COLLECTING MATERIALS

S tart collecting materials today—even if you haven't yet purchased land. No place to put them, you say? You probably have lots of free storage space at your disposal. Ask your parents or relatives if they don't mind lending a little extra yard space. Because, trust me, if the chance arises to pick up some free or cheap building materials, you don't want to pass it up.

Certain items, such as cabinets, interior wood trim, carpeting, and doors must stay dry. However, tiles, windows, toilets, bathtubs, and roofing can all be stored outside. Framing lumber can be stored outside as well but remember to put spacers between each layer of lumber so that air circulates around each piece. Cover this with a makeshift roof— old plywood, corrugated tin, or plastic will all do—and be careful not to wrap the lumber in a way that traps moisture inside.

It's true—friends may think you've become some sort of crazy packrat, but when they laugh, remind yourself that you're going to own your dream house, free and clear. And soon enough, you'll be the one laughing. . . all the way to the bank.

AVOID CUSTOM ITEMS

When building a home, many people start off by hiring an architect, which, indeed, is a good way to start. However, architects, often more interested in aesthetics than budget constraints, like to order custom items from catalogs. This means that the average owner builder is going to end up spending a great deal more than standard retail for his or her materials.

Even if you don't buy materials at auction sales or use salvaged materials, you can save a great deal of money by purchasing standard items at retail stores. Keep an eye out for sales. Those high-design, one-of-a-kind fixtures often get reduced.

Even the very resourceful builder will end up buying many of the materials for his or her new home at retail stores. Make certain to let the salesmen know you're building a whole house. You should be able to get the same discount as a contractor. Ask for it!

AUCTION SALES

Auctions have been around for thousands of years and are held in every part of the world. They're probably the cheapest place to buy building materials. However, you've got to have a little savvy about the whole process to make sure you get the best deal possible. When looking over the lots, carry a calculator and figure out the price you're willing to pay. Take into account that you will have to move what you buy and that you'll have a limited amount of time to remove your merchandise from the premises, usually two or three days.

I try not to pay more than 50 percent of the best price elsewhere. Generally, I pay only 20 to 30 percent of retail. From time to time, you'll see materials being sold for as little as 10 cents on the retail dollar.

There seems to be a certain mystique about auctions. Many people have preconceived notions that you have to be in an elite group to attend auctions. But, hey, we're not talking about *Sotheby's* or *Christie's*; we're talking about auctions for building materials. Auctions are for everybody. There're no "closed shop" auctions, and most people attending buy for their businesses and personal use.

GET THERE EARLY

When you attend an auction sale, make sure you have enough time to check things out. If possible, try to go the day before, but if you can't, then try to go early on the morning of the sale. Auctioneers usually allow you to preview at least two hours before the sale begins. They probably won't let you use their phone, so you'll have to use a public phone to check prices (unless you have a cell phone).

Get your bidder's number before you leave. You may have a long time to wait before you have to bid, especially if the item you want is a high number. A good auctioneer will sell about 100 to 125 lots an hour.

A 600 hundred lot auction will last six or seven hours. Very often the best bargains will be found at the end of the day because many buyers will have left by then. Try to stick around.

MY FIRST AUCTION

For some time, I'd heard other people boasting about their great buys at auction sales. I decided to attend one in hope of finding some materials for my house.

So it was that I went to my first auction, a fantastic, huge two-day event. I had no idea what I would find there. I ended up buying one lot of extension ladders (approximately twelve) for $4 each, one lot of twenty heavy-duty extension cords for $5 dollars each, an angle grinder for $25 dollars (still running great), a few boxes of nails for a dollar a box, and finally a load of plywood and lumber—including twenty sheets of new 5/8-inch plywood, a number of 12-foot long 8×10-inch beams, and some short pieces of telephone poles, which were later used in our retaining wall—all for $10. . . . Then, I started worrying about how I would get the whole pile home.

BUY A TRUCK

I bought a big truck for $700 with an 8×16-foot flatbed, which came in handy when I bought a backhoe at another auction. It was also invaluable when I bought a steel-building frame (story to follow). In the first few months of owning my truck, it more than paid for itself.

However, I wouldn't buy another if I didn't have to because they're difficult to insure, expensive to fix, and costly to run. I ended up selling it for $900 a year after I bought it.

If you plan on buying your materials at auction sales or in other unconventional ways, buy yourself a full-sized pickup truck with a trailer hitch or panel truck. Otherwise, you'll have the added expense of renting a truck each time you need to move building materials.

THE CHEAPEST WAY TO BUY MATERIALS

After my first auction, I was hooked! I decided that this was definitely the way to buy materials for my house and kept going to auctions whenever I could.

Once, I drove a hundred miles to an auction at a bankrupt tile company. There were several large lots of Brazilian tile neatly stacked together. I rummaged around and found another large lot of glazed ceramic Brazilian tile mixed with smaller quantities of plain beige and brown domestic tile, probably left over from other jobs. The neatly stacked lots fetched handsome prices. When the auctioneer finally got around to the disorganized lot, which I had my eye on, it sold for fifty dollars—to guess who?

I don't think anybody realized that there were more than 600 square feet of Brazilian tile in this lot. After buying some other tile recently at retail, I'm quite certain that my tile would sell for at least $4 or $5 a square foot! Which means that I bought $3,000 dollars worth of tile for $50 !

At another auction, I paid $1,800 for a container of lumber, plywood, and plywood siding. The value for the total lot was about $3,500. By my own rule of not paying more than half of retail, I paid too much. But since I needed the siding for my house, I thought what the heck?

You're not always going to find things for 10 cents on the dollar. When it's something you need, its value increases. Bid a little more for it—just don't get carried away.

THE STEEL BUILDING

When I made up my mind to build my house, the most difficult thing to decide was what I wanted it to look like. The materials I found at auction helped me to establish a vision for my home.

At one auction, I came upon a lot of big timbers, used for constructing new freeways and overpasses. They triggered the idea for a timber-frame house built with steel connectors instead of the traditional mortise and tenon joints. Unfortunately, someone wanted this lot more than I did, so he ended up getting the big timbers. Nevertheless, a seed of a notion had been planted.

One day I saw a newspaper advertisement for a big two-day auction at a railroad equipment and repair facility. Since I had planned to build retaining walls using steel posts and railroad ties, I knew that it would be worthwhile for me to attend. I saw the steel building on the first day. It was lying in pieces on the ground like a huge *Erector Set*. I had no idea how I would end up putting the pieces back together, but I was certain that I'd found my house.

It was a slow, arduous auction—more than 1500 lots in all—and to make matters worse, it was a very hot day. I bought steel railroad track and railroad ties for half the normal price. By late afternoon on the second day, most of the buyers had already gone. It was all left to just a few bidders, most of whom were scrap-metal dealers. We moved slowly around a huge yard, bidding on lot after lot. Everyone wanted to go home, especially the auctioneer.

We eventually came to the steel building. I had already decided what I would pay for it. But I was a little wet under the collar and started the bidding at the opening bid: $500. These days, if no one else makes a first bid, I'll make a lower bid than what the auctioneer first calls (in this case: $300). Anyway, only one other person bid against me, and it took him a long time to decide. He was probably as worried as I was about moving it. I bought the steel building for $900—I had been willing to go as high as $2,000.

Fig. 12. opposite page. The $900 steel building was to become our dream home.

An interesting bonus came along with the steel building. Since everyone got their materials out of the yard before me, the auctioneer allowed me to "clean up the yard." I was more than happy to offer a hand in gathering up free railroad ties and extra lengths of steel rail. The company vacating the auction site had paid an extra month's rent, so I was able to haul my loot in a leisurely manner.

LOOK PAST THE DIRT

When you're at an auction, try to look past the dirt. I once attended an auction at a kitchen and bath showroom. They sold a top name brand two-compartment ceramic sink for $75. Later out in the warehouse at the same auction, I bought an identical model for $5 because it was covered with dust.

OTHER AUCTIONS

The I.R.S., police departments, U.S. Customs' and public administrators' offices all periodically hold auctions. At a police auction I once bought a perfectly good *Xerox* copy machine for $4. A piece of paper was wrapped around the developing spool and the glass was broken. It cost me $35 to get it fixed. It continues to run well to this day.

HOW TO FIND AUCTIONS SALES

Companies hold auctions for many reasons. Generally, an auction is the fastest way of obtaining cash. As far as I know, most, if not all auctions are advertised in local newspapers. Sunday is the preferred day to advertise. Some newspapers have a weekly section of "display" ads. In the *San Francisco Examiner* and *Los Angeles Times* you can find these display ads all grouped together on one page. Don't ignore an auction just because you don't find exactly what you need listed in its newspaper ad. You never know what you'll dig up once you're there.

Peruse your own newspaper and figure out where auction sales are advertised. Though auctioneers usually advertise well in advance and often more than once in the same place, I would check the paper as often as possible so as not to miss a sale.

I also recommend flipping through the yellow pages for auction companies. Call them up directly when you figure out which auctioneers deal in the sort of goods you need. Ask to be put on their mailing list so that you receive mailings about up-and-coming auction sales.

MAKE SURE IT WORKS

If you're buying vehicles or electric tools, make sure they work. It's basic: don't buy it if it doesn't work! Items bought at auction sales are "as is" and not returnable. (The exception would be if the auctioneer misrepresented an item to be in working order when it wasn't.)

Air tools, such as nail guns, can be excellent bargains at auctions—examine pneumatic tools for cracks and missing parts. I recently bought a framing nail gun in like-new condition for $175 (it would have retailed for $400).

On the other hand, I've had mixed results with electric tools. And so I've come to the conclusion that there's no substitute for new electric tools.

DON'T PAY TOO MUCH

As I've already mentioned, I try not to pay more than 50 percent of retail. Usually, I pay 30 percent or less. Sometimes I'll break my own rules and pay a little more, especially if I need whatever it is right away. You shouldn't pay much more than 50 percent of retail because you have to move whatever it is and store it too.

Sometimes something may not seem quite right—the bidding is going too high, consistently. This may mean that there's "a shill" at work. A shill is someone who drives up bids because he has merchandise on consignment, or worse yet, because he's employed by the auctioneer.

While this last scenario is both illegal and rare, there's nothing to prevent an individual from trying to force a better price on his consigned merchandise. The best defense is not to get carried away when bidding. Decide ahead of time what you're willing to pay and stick pretty close to it. Shills can be hard to spot.

HOW TO BID

The most important aspect of bidding is getting the auctioneer's attention—without causing a scene. Once you have it, the auctioneer will look you in the eye after each counter-bid to see if you're still interested.

I usually give a slight nod once I have the auctioneer's attention. I prefer not to let everyone know I'm bidding. If I'm no longer interested, I'll indicate this by shaking my head or discontinuing my eye contact with the auctioneer.

When the auctioneer says "Sold!" and you're the successful bidder, hold up your number or shout it out. Auctioneers selling 100 lots or more per hour get annoyed if they have to wait for you to hunt for your bidder's number.

MOVING YOUR STUFF

At the commencement of an auction, the auctioneer will announce the terms and conditions. Something like this will be stated at each auction: "Every item is sold as is, where is, without a warranty expressed or implied. Cash or cashiers check. All bills must be paid today. Removal is today, tomorrow, and the next day only."

If you don't have all the money with you, in many cases, you can leave a deposit and pay the balance before you remove your purchases. You should plan on how you're going to move your newly acquired goods ahead of time.

BUILDING WRECKING YARDS

Building wrecking yards—like auto wrecking yards—can be found all over the country. Steel, pipe, bathtubs, toilets, bricks, flooring, siding, and beams are all readily available. I've purchased doors and windows at such yards. They were just as good as new, and sometimes they *were* new.

The maple flooring in my house came from a wrecking company. Originally, my flooring had been in San Francisco's main post office, which was later turned into a shopping center. I wouldn't have been able to afford maple flooring new, but at 75 cents a square foot, I was able to buy enough to cover over 2,000 square feet.

You can find building wrecking yards in the yellow pages under "building materials, used." See what the building wrecking yard nearest you has available because they generally sell materials at 40 to 50 percent less than retail.

FIG. 13 CALDWELL
BUILDING WRECKERS
*I bought the maple
flooring for our
dream house at
Caldwell Building
Wreckers.*

SALVAGE MATERIALS AND RECYCLED MATERIALS: FREE STUFF

In case you haven't noticed, there's a tremendous amount of waste in the construction business—both in new construction and remodeling. If you live in a major metropolitan area, keep your eyes open for large remodeling jobs where they throw away the nice dry straight framing members (they replace them with metal studs). You'll probably also see debris boxes filled with perfectly good 2×4s and plywood. Since they're throwing them away, you should be able to take them (but do ask first). It makes great sense to use recycled building materials. You're giving something a second life, and that's good for the environment.

Once, I bought a lot of trim material at a lumber yard auction—about $3000 worth of trim for only $350. Nobody had thought of selling the rack that had held the trim separately, so I got it for free. I disassembled the rack, which consisted of 12-foot-long 2×4s—about forty pieces in all. The 2×4s were actually quite old, but they'd been stored inside. So in effect, they were as good as new. The forty pieces now have a new lease on life as the walls of my living room, which, by the way, has 12-foot ceilings.

SCAVENGING

Always keep your eyes open for old barns and other dilapidated buildings that need tearing down. Of course, there's always the farmer who tries to make a fast buck on the friendly wood scavenger, but not too long ago I saw an ad in my local paper that said, "Free barn wood, you remove." (fig. 14a) Barns can be a fantastic source for siding. If you get a chance to tear down or tear out the insides of a building, you'll need these tools:

Wrecking Adze
Steel Shaft Claw Hammer
Crow Bar
Sledge Hammers (Large and Small)
Nail Pullers
Respirator
Gloves
Long Pry Bar

Serious scavenging also requires an oxyacetylene torch. I bought mine for $50 dollars at an auction. It's small standing—about 3-feet high—and is easily moved. Use the oxyacetylene torch to cut steel, bolts, and other metal.

Most home builders don't know how to use steel, but you can easily transform steel beams into nailing members by bolting a piece of wood to one or both of its sides. Steel often goes cheap at auction sales because scrap-metal dealers are usually the only ones who want it.

Proper procedure for dismantling a building is to remove all windows and doors first, then to take apart all plumbing and electrical, and then to salvage all cabinets and trim. Once the inside is stripped of its valuable, reusable materials, you start tearing down. The order should be: roof, roof framing, siding, walls, flooring and joists.

While you work, imagine how salvaged materials will be reused in your house. Pull nails out, and carefully stack the wood members. Try to keep your work area as clean and neat as possible; make sure you have an up-to-date tetanus shot; and, definitely, wear thick-soled shoes, long sleeves, jeans and gloves.

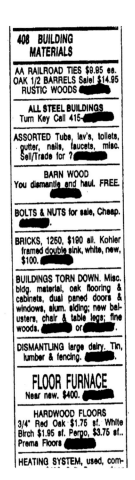

Fig. 14a. Classified ad.

Fig. 14. My oxyacetelene set.

NEWSPAPERS AND CLASSIFIEDS

Newspapers and classifieds can be good sources for odd bargains on building materials. I've seen ads for flooring, steel beams, wood beams, bathroom fixtures, kitchen fixtures, cabinets, windows, doors, tile, and much more.

Look in your local newspaper under "articles for sale" or "building materials." Both individuals and businesses sell materials this way. I bought a $3,000 central heating and air-conditioning unit for $400 through a newspaper ad. It works perfectly. I recommend that you know the value of the items you're looking at just to have a little extra bargaining power.

Fig. 15. This old barn is falling down, but it has yards of good wood that can be reused for siding.

GARAGE SALES

A student at one of my seminars told the class that neighborhood garage sales were his favorite place to look for bargains on materials. He asks people holding garage sales if they have any old wood, bricks, or other building materials that they would like taken off their hands. He often finds beautiful materials for next to nothing.

CONCLUSION.

Think of buying materials at a discount as a game, like finding the pieces to a very big puzzle. It can be exciting when unusual materials inspire unconventional building solutions.

Perhaps you worry that buying materials at a discount or at auction could somehow diminish the quality and craftsmanship of your dream house. I don't suggest buying something shoddy because it's cheap, nor do I think you should settle for a hodgepodge house. What I do suggest is figuring out ways to buy lasting, high quality materials—Italian marble, Brazilian tiles, maple floors, steel—at less than retail. By coupling imagination and interesting materials, you'll build a unique structure filled with unusual details that caters specifically to your aesthetics.

Next to land, materials will be the second largest expense in your dream house. You'll save tremendously if you try to buy as many materials as possible at some fraction of retail. Be patient. You won't find a bargain on everything in one day. But if you take your time and start collecting materials slowly, you'll be able to buy most of what you need at a discount. Remind yourself that this is a five-year plan to build and own your dream house free and clear. To make this happen, you don't want to cut corners on quality or craftsmanship. Patient collecting will reward you with the house you want at the price you want.

But if you are not able to find any materials in these unconventional ways, don't worry! In the next chapters I will tell you about three more money saving tactics.

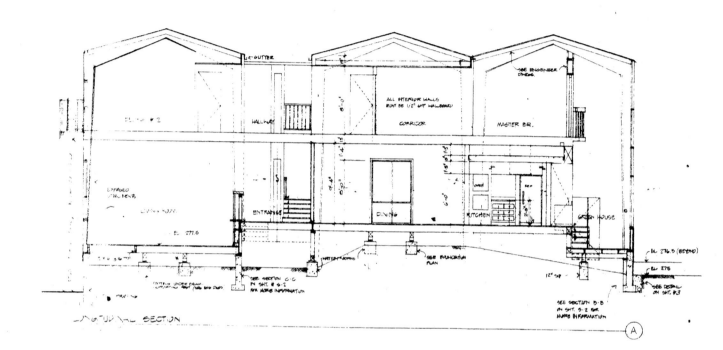

LONGITUDINAL SECTION

5   DESIGN

K I.S.S.: Keep It Simple Stupid. Unless you already have experience, let this motto govern the design of your home, whatever it may be—a log cabin, a lightweight wood frame construction, an adobe villa, or a timber frame house—try to keep it simple. And by *simple*, I mean that you design something that can be built in manageable stages. Thus, building your dream house will never become an overwhelming or daunting task.

All houses incorporate windows, doors, kitchen appliances, cabinets, bathtubs, toilets, plumbing fixtures, electrical fixtures, flooring. . . . So why do we choose one kind of construction over another? Aesthetics, economy, availability of materials, and ease of construction are all considerations.

I suggest that you design your home around the materials you've already collected—much in the way I was inspired by the steel building. Of course, if you hire a knowledgeable person to help you (see chapter seven) or you already have experience in one particular type of building, then this should also influence the design of your house.

By keeping the design of your new house simple, the construction will go smoothly and according to your best expectations.

REMODELING VERSUS NEW CONSTRUCTION

I don't suggest buying a house to remodel. You never find exactly what you want in a house originally built according to someone else's specifications. Remodeling is a dirty, unhealthy job, especially if you're living in the house while you do the work. More often than not remodeling ends

Our life is frittered away by detail. . . simplify, simplify.

HENRY DAVID THOREAU

up being more expensive than just starting from the ground up. The headaches from someone else's construction are analogous to those of buying a used car. But if you do find a great buy on a fixer-upper that you can't resist, I would advise remodeling before you move in.

HIRING AN ARCHITECT

Before building my dream house, I was so inexperienced I couldn't even read a blueprint. My wife and I definitely needed to hire an architect. You might want to hire an architect too, especially if you plan on doing something a little unusual.

Hire an architect who makes you feel comfortable. Agree on the price you're going to pay in writing before he or she begins drawing your house plans. This should be a fixed price because you don't want any surprises later. Though if your original vision changes, you will have to pay more. Architects expect to do a certain amount of work for an agreed price. It's wise to put an addendum into the contract so both you and the architect understand how much those changes will cost—it will almost certainly be an hourly rate.

It's much easier to erase a little mistake off the blueprints than to repour a foundation, so solve design problems before you start to build.

COMPUTER-AIDED DESIGN

Architect:
One who drafts
a plan of your
house, and plans
a draft of your
money.
AMBROSE BIERCE

On page 67-69, you can see the simple house I designed myself, using *Broderbund's 3-D Home Architect® Edition 2,* one of the many computer software programs that are now available for such purposes. The whole process took a few hours. Although some counties do require an architect's stamp on all your plans, in many places plans like these— generated with a computer at home and a book of details (please see page 74)— should more than satisfy the building department.

Here are a few criteria for selecting a program to design your house: (1) It should be capable of generating a floor plan showing elevations to scale, either 1/8- or 1/4-inch. (2) It should give a three-dimensional perspective of your new home, which makes it easier for you to envision the inside your new home before it's built. (3) It must be able to design a roof. (4) It must be capable of providing an itemized list of all the materials needed for building.

But again, if you're building your first house, you'll probably want to ask for some sort of professional help from a builder, an architect, a drafts-person, or perhaps, even another veteran owner builder.

THE ROOF

You might want a big house, or a small house, or a long house—and that's fine because any of these would still follow the rule of keeping it simple. But whatever else you want, please control the urge to build a house with seven gables.

The roof is generally the most difficult part of the house to build. Save yourself a big headache and keep the roof simple. My preferences for roof design are the shed roof, the flat roof, and the gable roof—all can be easily assembled by the lone owner builder.

The shed roof is, without a doubt, the easiest to construct. The design of the shed roof is exceptionally versatile and can be built from a number of different roofing materials, including composition shingles, roll roofing, metal, and tile.

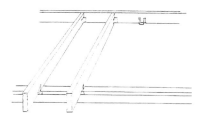

Fig. 18a. A gable roof with a ridge beam. The cap plate has been ripped to the angle of slope. Note the metal fastener.

The flat roof has a slope of not less than 1/4 inch per foot (for water drainage). This type of roof is traditionally made of tar and gravel; however, there are now products similar to roll roofing. These adhere to the previous course when heated with a propane torch. I used this product on our guest house and found it fairly easy to put down. If you're planning to build a flat roof, I'd ask your roofing supplier about this material.

The Gable roof is the most challenging of the three. I like to build this roof using a ridge beam, which is very easy to erect. Once the ridge beam is up, the rafters are simply positioned and fastened. Traditionally, rafters used in this manner are notched to fit snugly on the top of the wall (called a bird's mouth cut).

You can, however, employ a much more simple method, providing you have a table saw. To save labor, rip the top plate of the wall, also known as a cap plate, to the angle of slope. If you use this method, you must use a good fastener such as a Simpson strong tie (fig. 18a).

Roof trusses are another method for building a gable roof. You can build your own trusses or order them from a truss manufacturer. Though trusses can be quite beautiful, they are awkward to handle and require at least two people for raising. I have raised them single-handedly, but it's quite difficult compared to the relative ease of installing rafters.

ENTRANCE

It is very unpleasant to enter directly into the living areas. You should always have an entrance area, a small space to hang coats or remove boots, before entering the main house. A formal entrance preserves the privacy of the house's residents.

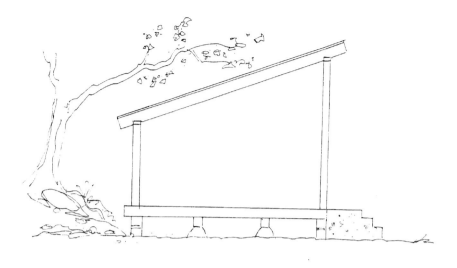

Fig. 16. A shed roof.

Fig. 17. A flat roof.

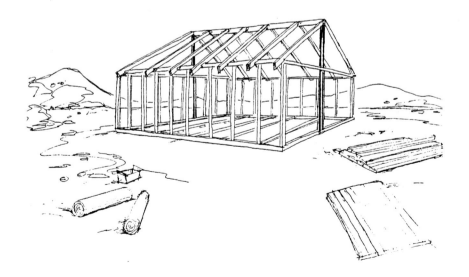

Fig. 18. A gable roof.

SAFETY

By building your own house, you become keenly aware of every possible danger built into it. You should be conscious of both immediate physical dangers (e.g. stairs) and unseen dangers from toxic substances.

Until recently, people gave little thought to the many poisonous and carcinogenic products used in building—hazardous substances can be found in paint, synthetic carpet, old plumbing pipes, and forced-air heating systems. If you build your own home, you'll be able to scrutinize every item used in the construction process.

Stairs are probably the greatest physical menace in a house, and the elderly are at greatest risk. You may want to consider building a one-story house simply because it's significantly easier to build than a multi-story house. I learned this the hard way. I fell from my second story while I was attaching the plywood subflooring. Fortunately, I landed on my feet (literally), however my ankles were badly bruised.

It's true that the material costs of building a one story house are slightly higher, but in the bottom-line they're outweighed by labor savings. If you build a multi-story house, you'll be surprised by how much time you spend hauling materials up and down.

ORIENTATION

Orient your house to take advantage of sun and views. In the northern hemisphere, a southern orientation is usually preferred for capturing the most sunlight year-round. Tract-developers, concerned with speed of building and bottom-line profits, spend little time worrying about which way an individual house faces.

DESIGNING YOUR DREAM HOUSE FOR A $100

As I mentioned earlier, I used *Broderbund's 3-D Home Architect® Edition 2* to design a simple, easy-to-build home. Until recently, nothing like this technological tool was available, and so you had no choice but to pay an architect hundreds of dollars to do a three-dimensional view.

The dialog between client and architect is about as intimate as any conversation you can have, because when you're talking about building a house, you're talking about dreams.
ROBERT A.M. STERNS
The New York Times
Jan 13, 1985.

Fig. 19. An artist's drawing of the example house.

The best way to realize the pleasure of feeling rich is to live in a smaller house than your means would entitle you to.
EDWARD CLARK
The Story Of My Life

Today, you can view your new home with the touch of a button from any angle and print out the result. The floor plan in figure 22 has been modified to fit the page, so it's not to any particular scale. You can, however, specify either 1/8- or 1/4-inch scale and print to these specifications. Figures 20 and 21 show three-dimensional views. Figure 19 shows the finished house. With this program, I could have chosen any of the three roof styles already discussed.

A FEW COMMENTS ON THE HOUSE I DESIGNED

I wanted to design a simple, pleasant house that I could build by myself. I also wanted to show how much a software program can do and how *Broderbund's 3-D Home Architect® Edition 2* generates a list of materials (see pages 70-74).

One thing I don't like about this program is that there are several items missing from the list of materials; therefore, an allowance should be made for rough plumbing and electrical, floor and ceiling insulation,

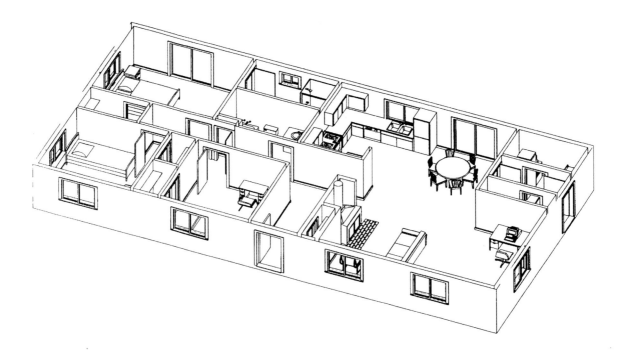

Fig. 20. 3-D view from the front.

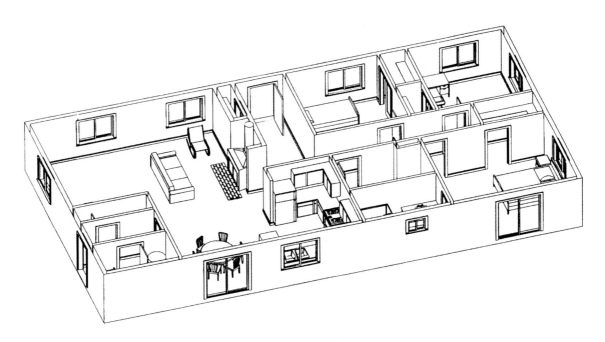

Fig. 21. 3-D view from the rear.

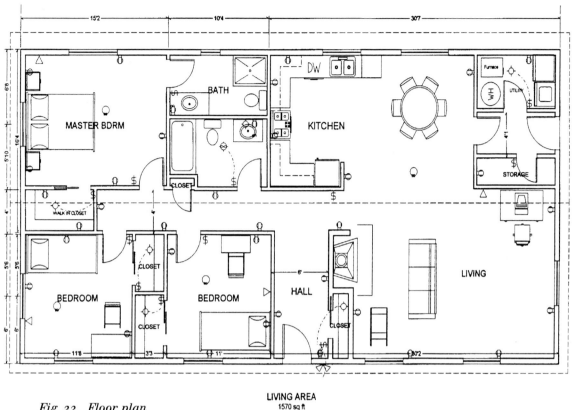

Fig. 22. Floor plan.

LIVING AREA
1570 sq ft

paint, and nails. Also, you can't specify tongue and groove hardwood flooring, which is competitively priced, fairly easy to install, and aesthetically pleasing.

All in all, I have to say that this program does a pretty good job, especially considering its price (under $70). New materials to build this three bedroom house cost less than $30,000. You'll notice that I found a few items in the materials list on sale. All items were available between the two home improvement centers near my home. So even if you're not able to buy any materials for less than retail, you can still build a very nice house for a fraction of what you would expect to pay if someone else built it.

	ID	l	Size	Item	oun	xtr	Un	Quantity	Price	Tot. Cost	Comment
1											
2	GN1	1		heated wall area	1091	0	sqft	1			
3	GN2	1		heated glass area	199	0	sqft	1			
4	GN3	1		heated door area	40	0	sqft	1			
5				Subtotal:							
6											
7	FO1	0	4x18"h	concrete found. wall	4.63	0	cuyd	1	$84.50	$391.24	
8	FO2	0	2x6"	treated mud sill	223	0	ft	1	$0.75	$167.25	
9	FO3	0	1/2x6"	foam sill seal	223	0	ft	1			
10	FO4	0		foundation bolts	42	0		1	$0.50	$21.00	
11	FO5	0	12x10"h	concrete footing	6.86	0	cuyd	1	$84.50	$579.67	
12				Subtotal:						$1,159.16	
13											
14	SF6	1	4x8' sheets	plywood subfloor	49	0		1	$16.96	$831.04	
15	SF7	1	2x10" 16" OC	floor joists	1304	0	ft	1	$1.16	$1,512.64	
16	SF8	1	2x4" 16" OC	ceiling joists	1304	0	ft	1	$0.36	$469.44	
17				Subtotal:						$2,813.12	
18											
19	F1	1	2x6-91 1/2"	fir stud	185	0		1	$4.50	$832.50	
20	F2	1	2x6-16ft+	fir stud	564	0	ft	1	$0.57	$321.48	
21	F3	1	4x12"	door/window header	95	0	ft	1	$3.32	$315.40	
22	F4	1	2x4-91 1/2"	fir stud	172	0		1	$2.81	$483.32	
23	F5	1	2x4-16ft+	fir stud	538	0	ft	1	$0.36	$193.68	
24	F6	2	2x6-91 1/2"	fir stud	52	0		1	$4.50	$234.00	
25	F7	2	2x6-16ft+	fir stud	160	0	ft	1	$0.57	$91.20	
26				Subtotal:						$2,471.58	
27											
28	S1	1	4x8' sheets	ext. wall sheathing	47	0		1	$5.99	$281.53	
29	S2	1		exterior siding	1495	0	sqft	1	$0.50	$747.50	
30				Subtotal:						$1,029.03	
31											
32	EX1	1		exterior sill	50	0	ft	1	$0.85	$42.50	
33	EX2	1		ext. window casing	303	0	ft	1	$0.45	$136.35	

Materials list page 1.

	ID	l	Size	Item	oun	xtr	Un	Quantity	Price	Tot. Cost	Comment
34	EX3	1	1x6-36"	door threshold	2	0		1	$10.00	$20.00	
35	EX4	1		ext. door casing	35	0	ft	1	$0.45	$15.75	
36	EX5	1	6 in	ext. door jamb	35	0	ft	1	$4.11	$143.85	
37				Subtotal:						$358.45	
38											
39	R1	1		roofing material	1860	0	sqft	1	$0.25	$465.00	
40	R2	1	4x8' sheets	roof sheathing	59	0		1	$10.98	$647.82	
41	R3	1	2x6" 16" OC	rafters - fir	1575	0	ft	1	$0.57	$897.75	
42	R4	1	2x8"	gable fascia	64	0	ft	1	$0.85	$54.40	
43	R5	1	2x8"	eave fascia	116	0	ft	1	$0.85	$98.60	
44	R6	1		gutter	116	0	ft	1	$0.36	$41.76	
45				Subtotal:						$2,205.33	
46											
47	IN1	1	6x16x93" bats	wall insulation	103	0		1	$2.50	$257.50	
48											
49	FL1	1	12x12x1/4	tile floor	217	0		1	$2.00	$434.00	
50	FL2	1		carpet	1336	0	sqft	1	$1.74	$2,324.64	Red oak shorts
51				Subtotal:						$2,758.64	
52											
53	WB1	1	4x8'-1/2"	wp wall board	6	0		1	$6.40	$38.40	
54	WB2	1	4x8'-1/2"	wall board	219	0		1	$3.00	$657.00	
55				Subtotal:						$695.40	
56											
57	W1	1	60x80	alum. right sliding	1	0		1	$184.00	$184.00	
58	W2	1	24x24	alum. left sliding	1	0		1	$40.00	$40.00	
59	W3	1	48x48	alum. left sliding	5	0		1	$70.50	$352.50	On sale
60	W4	1	60x80	alum. left sliding	1	0		1	$184.00	$184.00	
61	W5	1	48x48	alum. right sliding	3	0		1	$70.50	$211.50	On sale
62				Subtotal:						$972.00	
63											
64	D1	1	36x80x1 3/4L	ext. hinged	1	0		1	$199.00	$199.00	
65	D2	1	36x80x1 3/8R	ext. hinged	1	0		1	$107.00	$107.00	
66	D3	1	30x80x1 3/8R	hinged	4	0		1	$35.00	$140.00	

Materials list page 2.

	ID	1	Size	Item	oun	xtr	Un	Quantity	Price	Tot. Cost	Comment
67	D4	1	26x80x1 3/8L	hinged	1	0		1	$32.00	$32.00	
68	D5	1	30x80x1 3/8	pocket	2	0		1	$65.00	$130.00	
69	D6	1	23x80x1 3/8L	hinged	1	0		1	$28.00	$28.00	
70	D7	1	30x80x1 3/8L	hinged	3	0		1	$35.00	$105.00	
71	D8	1	48x80	slider-panel	2	0		1	$58.00	$116.00	Mirror doors
72				Subtotal:						$857.00	
73											
74	C1	1	BD4836	bath base cab	1	0		1	$149.00	$149.00	
75	C2	1	PBD3636	bath base cab	1	0		1			
76	C3	1	UDR2484	util cab	1	0		1	$177.00	$177.00	
77	C4	1	BLBDR1836	base cab	1	0		1	$68.00	$68.00	
78	C5	1	BD3036	base cab	1	0		1	$101.00	$101.00	
79	C6	1	BDR1836	base cab	1	0		1	$68.00	$68.00	
80	C7	1	BS4836	base cab	1	0		1	$80.00	$80.00	
81	C8	1	32x12 in	shelf & rod	2	0		1	$7.50	$15.00	
82	C9	1	PBDR2436	base cab	1	0		1	$78.00	$78.00	
83	C10	1	BLPBDR2336	base cab	1	0		1	$78.00	$78.00	
84	C11	1	BDR2436	base cab	1	0		1	$78.00	$78.00	
85	C12	1	37x12 in	shelf & rod	1	0		1	$7.50	$7.50	
86	C14	1	WR2430	wall cab	2	0		1	$51.00	$102.00	
87	C15	1	BLWR2730	wall cab	1	0		1	$54.00	$54.00	
88	C16	1	W3030	wall cab	2	0		1	$56.00	$112.00	
89	C20	1	BLWR1330	wall cab	1	0		1	$36.00	$36.00	
90	C22	1	WR1830	wall cab	1	0		1	$46.00	$46.00	
91	C23	1	61x12 in	shelf & rod	1	0		1	$12.50	$12.50	
92	C24	1	67x12 in	shelf & rod	1	0		1	$15.00	$15.00	
93	C25	1	84x12 in	shelf & rod	1	0		1	$17.50	$17.50	
94				Subtotal:						$1,294.50	
95											
96	T1	1	1x4-16ft+	window apron	50	0	ft	1	$0.69	$34.50	
97	T2	1	1x2-16ft+	sill	50	0	ft	1	$0.43	$21.50	
98	T3	1	1x4-16ft+	interior casing	633	0	ft	1	$0.31	$196.23	On sale
99	T4	1	4 in	interior jamb	206	0	ft	1	$1.40	$288.40	
100	T5	1	6 in	interior jamb	17	0	ft	1	$1.80	$30.60	
101	T6	1	1x6-16ft+	base moulding	568	0	ft	1	$0.79	$448.72	

Materials list page 3.

	ID	l	Size	Item	oun	xtr	Un	Quantity	Price	Tot. Cost	Comment
102				Subtotal:						$1,019.95	
103											
104	FX1	1	42W33D	33x42" Shower	1	0		1	$199.00	$199.00	
105	FX2	1	30W28D	Standard Toilet	2	0		1	$79.00	$158.00	
106	FX3	1	20W16D	Oval	2	0		1	$49.00	$98.00	
107	FX4	1	60W32D	Standard 60R" Tub	1	0		1	$215.00	$215.00	
108	FX5	1	25W25D	Gas Water Heater	1	0		1	$233.00	$233.00	Inc vent
109	FX6	1	48W25D	48x29" Std Fireplace	1	0		1	$1,200.00	$1,200.00	Allowance
110	FX7	1	31W21D	32" Double Kit. Sink	1	0		1	$138.00	$138.00	
111				Subtotal:						$2,241.00	
112											
113	A1	1	28W28D	Gas Dryer	1	0		1	$299.00	$299.00	
114	A2	1	27W28D	Washer	1	0		1	$349.00	$349.00	
115	A3	1	21W24D	Dish- washer	1	0		1	$261.00	$261.00	
116	A4	1	31W25D	30" Gas Range	1	0		1	$479.00	$479.00	Inc hood
117	A5	1	30W30D	30"R Ref/ Freezer	1	0		1	$469.00	$469.00	
118	A6	1	22W29D	20" Lg Gas Furnace	1	0		1	$900.50	$900.50	Inc duct & vent
119				Subtotal:						$2,757.50	
120											
121	FU1	1	85W35D	7' Sofa	1	0		1			
122	FU2	1	21W40D	Rocking Chair	1	0		1			
123	FU3	1	68W70D	48" Diam. Seat 5-6	1	0		1			
124	FU4	1	54W33D	54"x32" Contemporary	1	0		1			
125	FU5	1	64W84D	Queen Bed Contemp.	1	0		1			
126	FU6	1	42W88D	Contemp. Long Twin	2	0		1			
127	FU7	1	36W18D	36"x18" Traditional	1	0		1			
128	FU8	1	48W24D	48"x24" Contemporary	1	0		1			
129	FU9	1	26W26D	Swivel Chair	1	0		1			
130	FU10	1	20W27D	Side Chair	2	0		1			
131	FU11	1	20W18D	2 Drawer	2	0		1			
132				Subtotal:							
133											
134	E1	1	wall mount	110V duplex outlet	35	0		1	$0.39	$13.65	
135	E2	1	wall mount	Telephone Jack	4	0		1	$1.99	$7.96	

Materials list page 4.

	ID	l	Size	Item	oun	xtr	Un	Quantity	Price	Tot. Cost	Comment
136	E3	1	wall mount	Outdoor Dual Spot	1	0		1	$18.88	$18.88	
137	E4	1	wall mount	wall switch	13	0		1	$0.49	$6.37	
138	E5	1	wall mount	std light	2	0		1	$19.99	$39.98	
139	E6	1	wall mount	110V GFI Dup. Outlet	2	0		1	$7.99	$15.98	
140	E7	1	ceiling mount	Smoke Detector	4	0		1	$19.80	$79.20	
141	E8	1	ceiling mount	Rec. Light/Fan	1	0		1	$117.00	$117.00	
142	E9	1	ceiling mount	Surface Mnt. Fixture	4	0		1	$39.95	$159.80	
143				Subtotal:						$458.82	
144											
145	H1	1		bath cab door handle	4	0		1			N/C
146	H2	1		cab door handle	20	0		1			"
147	H3	1		cab drawer handle	6	0		1			"
148				Subtotal:							
149											
150				TOTAL:						$23,348.98	

Materials list page 5.

THE REST OF THE STORY

In order to satisfy the building department, your plans must include "details" of your foundation, walls, and roof. You must also be able to explain how you intend to connect all the structural elements together.

It's possible to buy a book of details, copy what you need, and then incorporate them into your plans. Details are drawn to scale. I frequently refer to *The DETAIL Book* (see bibliography).

With the publisher's permission, I've included several details from this book in the next few pages. Currently, the only other book available for such purposes is *The Graphic Guide to Frame Constrution* (Tauton Press).

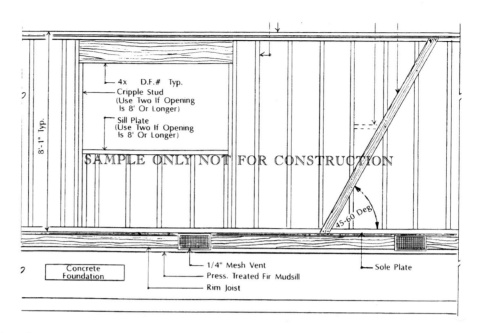

4x   D.F.# Typ.
Cripple Stud
(Use Two If Opening
Is 8' Or Longer)
Sill Plate
(Use Two If Opening
Is 8' Or Longer)

SAMPLE ONLY NOT FOR CONSTRUCTION

8'-1" Typ.

45-60 Deg.

Concrete
Foundation

1/4" Mesh Vent
Press. Treated Fir Mudsill
Rim Joist

Sole Plate

*Fig. 23. Typical stud
wall construction.*

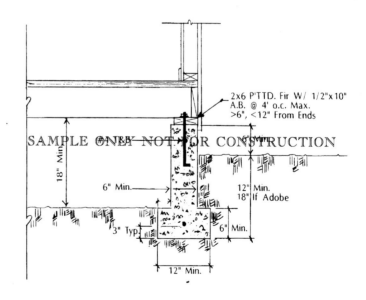

2x6 P'TTD. Fir W/ 1/2"x10"
A.B. @ 4' o.c. Max.
>6", <12" From Ends

SAMPLE ONLY NOT FOR CONSTRUCTION

18" Min.

6" Min.

12" Min.
18" If Adobe

6" Min.

3" Typ.

12" Min.

*Fig. 24. Single story
perimeter foundation.*

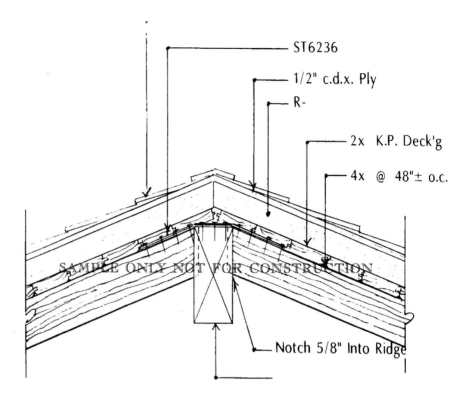

ST6236

1/2" c.d.x. Ply

R-

2x K.P. Deck'g

4x @ 48"± o.c.

SAMPLE ONLY NOT FOR CONSTRUCTION

Notch 5/8" Into Ridge

*Fig. 25. Exposed beam
ceiling at ridge.*

COMPLETING A SET OF PLANS

*Few rich people own
their property.
Their property
owns them.*
ROBERT GREEN INGERSOLL
*Address to The McKinley
League 1896*

The required amount of detail on your plans depends on your building department. Get a list of what's required.

In addition to the house plans, you'll be required to provide a vicinity map, a location map, and a site plan. A vicinity map shows where your lot is in relation to nearby streets and other landmarks. A location map indicates where your project is located in the county. Both vicinity and location maps are easily acquired from county offices or other sources. However, a site plan, showing exactly where your house will be built on your property, will take a little work in drawing up. You may need the help of a surveyor, especially if your property is hilly or contains other unusual geographical features.

Energy calculations may also be required. These calculations demonstrate compliance to energy codes and should take windows, insulation, and the heating system into account. There are many companies that specialize in this service, and the cost is usually minimal.

CONCLUSION

Dream house is, of course, a relative term—your dream house may be a cottage or it may be a mansion. It is self evident that your dreams will somehow have to be controlled by your pocket book if they are to become a reality. If you live in a rented apartment, any home that you design, build, and own free and clear should be a dream house. The main thing to remember is whatever you design let it be easy to build. Even if you buy land and materials at retail, you can still save a bundle by designing a house that is easy to build.

6 TEMPORARY DWELLINGS

There are a number of options for living on your land and paying no rent while you build. Obviously, paying no rent can save you a great deal of money, which in turn can be put toward the cost of your dream house. Following are four options: (1) build a small guest house, (2) build your house in stages, (3) build a small house that will be converted into a detached garage, or (4) buy a mobile home.

My family and I lived in a mobile home while we built, and it was actually much nicer than we had anticipated. The great thing about a mobile home is that it comes equipped with a kitchen, storage, bathroom and a number of other amenities. This means you don't get bogged down with building a comfortable temporary dwelling, but, instead, can focus on the construction of your dream house right away.

In the United States, thousands of older mobile homes are available for under $5,000 each. You can buy one, live in it while you build, and then sell it to recover most if not all of your investment.

Many counties allow you to live in a mobile home on your land while you build. We had no trouble getting a permit in Marin County for our mobile home. Living there while we built saved us over $20,000 in rent—all of which we were able to apply to materials and labor. To get your permit, you'll be required to hook up to the sewer or install a septic system. You'll also need a source of drinking water and probably want to install electricity and a telephone.

MOBILE HOME PERMITS

Some counties allow long-term placement of mobile homes. You may be required to put in a regular foundation for such a permit. Living in a mobile home with a permanent placement permit could be used as a way to save for quite a while until you have enough money to build your dream house. When you finally did get your house built, you could probably keep the mobile home as a guest house. Sometimes, however, you'll be forced to get rid of it—in which case, you could use the old mobile home's foundation for another building, like a garage or guest cottage.

Many counties only allow temporary placement of mobile homes, usually for one year, conditional upon obtaining the required use permit. However, an additional one-year extension is often available. I doubt that authorities would try to evict you from your own land, especially if you're making progress on your home.

Some private developments may not allow you to live on your land while you build. Before you buy, be sure to check the covenants, conditions and restrictions (C.C. and R's.).

PURCHASING A MOBILE HOME

I found out about my mobile home in the newspaper under "the mobile homes for sale" category of the classifieds. There were about six of them listed for a closed-bid auction (closed bid means you can only make one written bid on each item) being held by a public administrator's office in a nearby county.

The homes were open for inspection. I drove approximately a hundred miles to inspect all of them, and I made three bids for my first, second, and third choice. Though there were homes in better condition, my first choice—since I have a family—was the biggest. It was 12×60 feet with an additional 80 square feet, called a pop-out, used for a dining room, had two bedrooms and two bathrooms and was the one we ended up with. It served us well for two years.

I bid $3,050. Perhaps, I could have gotten it for less, but I was getting pretty desperate. My building permit had been approved, and I was paying a $1,000 a month rent since selling my last home a few months earlier. It cost $300 to move it onto my site from the mobile-home park.

I did a little work on it, such as replacing the worn carpeting with new linoleum. It was very livable after I finished fixing it up—almost as nice as our rented house had been.

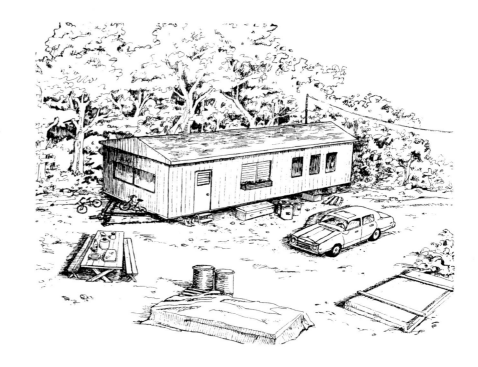

Fig. 26. Our family lived comfortably in a mobile home similar to this one for two years.

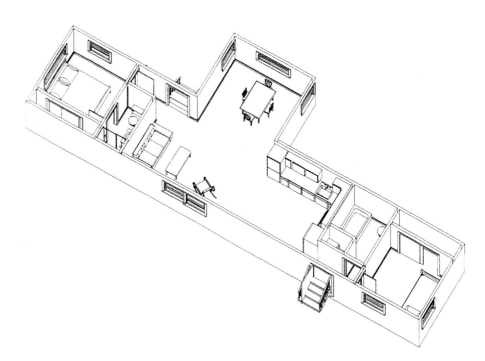

Fig. 27. This 3-D view of our Mobile home was created with Broderbund's 3-D Home Architect ®. See chapter 5

Fig. 28. Emily, Liz, and a friend studying hard in the pop-out.

When it came time to sell, I advertised our trusty mobile home in the newspaper for $5,650. Quite a few people enquired. I ended up selling it for $5,250 to a couple getting ready to build their own home.

MOVING A MOBILE HOME

Call two or three mobile home movers and get bids on how much they charge to move mobile homes. Be sure you get the mobile home ready to move. This includes folding and securing any pop-outs, removing any foundation blocks, and disconnecting utilities. You may also have to put air in the tires. The driver who moved mine had a compressor on his truck, so he put air in the one flat tire. Do not put anything inside the mobile home, especially concrete foundation blocks. It's very important to keep the home as light as possible to avoid any damage during moving.

SETTING UP A MOBILE HOME

If you have a temporary dwelling permit, setting up your mobile home should be fairly easy, because you won't be required to build a permanent foundation.

For a temporary foundation, you can use railroad ties, concrete blocks, or other sturdy beams of steel or wood. In any case, make sure you have plenty of support. You'll need support at least every 8 feet over the entire length of your mobile home. Also, your mobile home must be completely level or you will have trouble closing the doors.

I wouldn't use anything less than an eight-ton jack. You should have a good sturdy base for your jack, such as a square block of wood, to ensure it doesn't sink into the ground. In addition, make sure you have a good support under the hitch. Since we were on a slope, we had to jack up the back of our mobile home over 4 feet. It's important to remember the elementary laws of physics at all times of the jacking process. For each action you'll get an equal and opposite reaction—therefore, jacking up one side will cause the opposite side to come down.

One more reminder of the obvious: block the wheels before you begin! But don't block your mobile home so low that you can't crawl underneath (in case you need to fix something, such as a broken sewer line). Use the space under your mobile home for dry storage.

SEWER LINES

The county required that we hook up to the local sewer line. We were able to live in our mobile home with a rented portable outhouse until this task was completed, but I'm the first to admit this was pretty inconvenient, especially with a five-year-old.

You can see the sewer or drain lines under your mobile home. They probably will be 3-inch ABS pipe. The main thing you have to remember is that for proper drainage you must have a slope of no less than 1/4 of an inch per foot. The connections must be tight (the county health department won't allow any leaks). If you live in a cold climate, you want to be careful because a leaky system can freeze, causing pipes to burst and sometimes your toilet to jump up from the floor!

In my particular case of putting in the permanent sewer line the first job was to dig a trench—the last 45 feet of which needed a jack hammer. My problems were twofold. First, I had to be careful not to dig into a gas line, water line, or anything that may have been buried under the street. This problem was fairly easy to solve because the utility company sent out people (free of charge) to mark the exact position of all underground utilities. My second problem was that I had created an open trench in the street. This meant that I had to cover the trench at night and put up a flashing light and a couple of those sandwich-board barriers to warn people. By parking my backhoe over the end of my driveway, I covered one portion of the trench. Because we are on a cul-de-sac, I was able to lay sheets of 11/8-inch plywood on the street portion.

In order to hook up directly into the sewer, the sanitary district obliged us to use castiron pipe underground. (They have since changed the requirement to allow the use of plastic ABS pipe which is about one quarter the price). In any case, the installation of the castiron sewer pipe was quite simple. It was just a matter of securing ten-foot lengths of 4-inch castiron pipe together with special fittings, called no-hub couplings (fig. 28a), which were then tightened with a small rachet-type torque wrench.

I placed ample sand beneath and above the pipes to protect them from large rocks. When filling in the trench, or backfilling, you are supposed to compact the fill in layers. Since my trench was off the side of the road, I simply filled in the trench and drove the backhoe up and down the length of the trench. I graded it by pulling a bit more earth over the top of the trench, repeating the process until it was thoroughly compacted.

Fig. 28a. A no-hub coupling

WATERLINES

We had a temporary water meter installed. The permanent water meter would have cost over $6,000 and the temporary meter was only $150. Of course, we had to buy the permanent water meter before our house was completed, but using the temporary meter gave us the time to raise money for the permanent one.

PVC pipe was used from the temporary water meter to the mobile home and positioned while I was installing the sewage lines. It was easy to glue the lengths of pipe together and cover them with the required foot and a half of earth.

ELECTRICAL SUPPLY

Our temporary electrical supply was grounded at the source. We also installed a second ground rod. Be very careful about grounding your

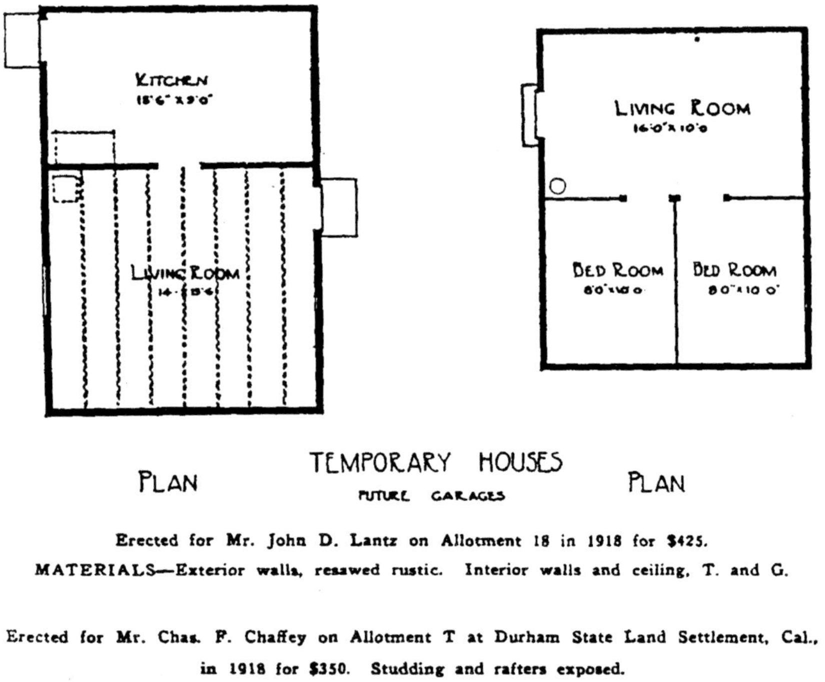
Kitchen
Living Room
Living Room
Bed Room
Bed Room
Plan
Temporary Houses
Future Garages
Plan
Erected for Mr. John D. Lantz on Allotment 18 in 1918 for $425.
MATERIALS—Exterior walls, resawed rustic. Interior walls and ceiling, T. and G.
Erected for Mr. Chas. F. Chaffey on Allotment T at Durham State Land Settlement, Cal., in 1918 for $350. Studding and rafters exposed.

Fig. 29. Temporary dwellings— future garages— pictured in a 1920 publication.

electrical supply. You're putting yourself and your family at risk of electrical shock if you don't ground properly.

Mobile homes usually have provisions for dual groundings—one to a waterline and one to a ground rod. However, if you use plastic pipe for your water supply as we did, it can't be used as a ground. I strongly recommend consulting a qualified electrician to help you properly put in a second grounding.

PROPANE VS. NATURAL GAS

Gas companies generally won't hook up your gas lines until you've completed your permanent residence. So it's more than likely that you'll have to use propane instead of natural gas while living in your mobile home.

If the previous owner of your mobile home was using natural gas, you'll have to make some modifications on your gas appliances. Your local propane dealer can advise you.

You should buy or rent two propane tanks with an automatic switch-over, so you don't run out right in the middle of taking a shower or cooking dinner. I bought two tanks at an auction for $50 each and paid $50 for the automatic switch-over.

OTHER OPTIONS

You may wish to consider other options for living on your land as you build. However, the main drawback to any option other than the mobile home is that you won't have a completed temporary dwelling immediately.

I think the most appealing option after the mobile home would be to build a small guest house with one or two bedrooms, a bathroom, a small living room, and a bare essentials kitchen.

Another viable option is to build your house in stages. This is what the first settlers did, and it worked very effectively for them. Rooms change function as the house expands. Thus, your first bedroom, built next to the kitchen, would eventually become a dining room.

The third option—if your county allows it—is to build a detached garage first. In Australia, where I lived for several years, it's very common to live in the garage while building. I'm sure people would live in mobile homes if they could, but unfortunately mobile homes are in short supply and very expensive. If you choose this third option, make sure to design and build your garage with no interior bearing walls. This way, you'll be able to divide the space to suit your needs while you're living there. Then, when you're ready to move into your finished house, you can easily remove the divisions without affecting the structural integrity of your garage.

MOVING INTO YOUR NEW HOUSE EARLY

You should be able to move into your house before it's completely finished. You'll need a bathroom and a cooking area. Make sure that you've accounted for all the hazardous situations, such as exposed wiring and unsecured second floor openings.

You'll find it difficult to work in partially completed rooms. I recommend completely finishing those areas you'll be living in day-to-day.

CONCLUSION

My family and I were doubtful that we could find a tolerable temporary solution for living on our land while we built. All the options seemed as if they would be inconvenient and uncomfortable. But honestly, we were very happy with our mobile home. Whoever designed it truly understood the practicalities of everyday life. We never felt cramped nor that our privacy suffered.

If you're not able to save by other methods (e.g., buying land and materials for less than retail), living on your land while you build is the surest way to save a bundle.

7 GETTING HELP

L abor accounts for at least 40 percent of the cost of building a new house. Of the major money saving ideas (buying land and materials for less than retail, designing for simplicity of construction, and living on your land while you build) directly engaging hands-on in the building process is probably the most significant thing you can do to realize the dream of owning your house free and clear in five years.

I believe that you must make a commitment from the very start to participate in the actual construction of your dream house. This doesn't mean that you have to build the entire house by yourself—quite to the contrary, I recommend hiring competent help. In fact, if there's anything you shouldn't try to skimp on, it's the quality of the people helping you to build your dream house. This doesn't necessarily entail hiring a contractor or professional construction workers. I suggest that you make a trade of services with another seasoned owner builder or that you find a retired builder who would enjoy working on your project part-time simply because he or she loves to build.

I want to emphasize, however, that you shouldn't walk off the job when you hire others. There are good reasons for participating—not only because you'll save on the cost of your own labor, but, more importantly, you'll maintain direction of the project. What I'm trying to say in euphemistic terms is that you *should* get help, but be wary of being "ripped-off" by those who aren't worth their pay. The best way to figure this out is to be on-site as much as possible. Trust me, things quickly go astray if you're not around to provide the necessary encouragement and wide-spanning vision for what is, after all, one of the most important projects of your life.

LEARNING TO BUILD

Before building my dream house, I'd never done any real construction—
I'd barely picked up a hammer for some minor remodeling. I decided
that it would be worth my while to get some additional building
experience.

I signed up for some short classes at the Owner-Builder Center in
Berkeley, California. Later, I signed up for a three-week, hands-on, live-
in building camp, which was held just north of San Francisco. About
thirty men and women attended the camp with me. The curriculum
included grading, layout, foundation, framing, roofing, plumbing,
electrical, and finishing an interior. We learned which tools to use for
what as well as how to build safely.

All in all, I have to say that the most substantial thing we learned was
how to think our way through a building problem. Unless it's tract-built,
every house is unique and will have unique problems for its construction.
No matter how strong you are, you can't build anything unless you can
stop at a critical juncture and think up a building solution. Honestly,
building a house is 90 percent brains and 10 percent brawn.

There are many ways to gain experience. I suggest that you go to a
workshop for owner builders or do some volunteer work. There are
many wonderful, charitable organizations that build houses all over this
country for those in need of decent shelter. Volunteering for such an
organization not only is a good deed in and of itself but also offers a
great opportunity to learn how to do construction. But whatever you
choose for gaining experience, I advise that you hire a competent helper
when you actually begin construction on your dream house.

My friend John, pictured on the front page of this chapter, retired
early from his job. He had a great deal of experience—he'd remodeled a
number of houses, moved an entire house in Canada, and worked with
steel. He worked with me on my dream house part-time three days a
week. I worked the rest of the week by myself. Working together side by
side, we raised my house, and over the course of the project, John
imparted a wealth of building knowledge that is now mine for good.

SCHOOLS FOR OWNER BUILDERS (LISTED BY STATE)

Here's a list of owner-builder schools and other helpful organizations. For additional courses, I would ask at your local schools and colleges to find out if they offer adult education. I teach a class—based on the principles outlined in this book—at a local high school's adult education program in Marin County, California.

ALASKA

ALASKA CRAFTSMAN HOME PROGRAM, INC.
900 Fireweed, #201
Anchorage, AK 99503-2509
(907) 258-2247

A.C.H.P. is an educational building industry alliance promoting energy efficient housing that is cost effective, healthier, and durable. They offer workshops on home building, airtightness, heating, and ventilation. They also provide manuals on all of the topics above.

ARIZONA

OUT ON BALE—BY MAIL
1037 E. Linden St.
Tucson, AZ 85719
(520) 624-1673

Hey! Straw is cheap—$3 to $4 a bale. Straw is environmentally friendly and a renewable resource. More and more, building departments are beginning to accept this technique. If you want to learn the basics of building straw-bale wall systems, this is the place. Out On Bale conducts hands-on classes in other states and even abroad. Videos and books are available.

CALIFORNIA

BUILDING EDUCATION CENTER
812 Page St.
Berkeley, CA 94710
(510) 525-7610

The Building Education Center is the successor to the Owner-Builder
Center where I learned the basics. Unfortunately, they're no longer
holding three-week long house-building camps, but they do have hands-
on classes in everything from drywall to plumbing.

RAMMED EARTH WORKS ASSOCIATION
1058 2nd Ave.
Napa, CA 94558
(707) 224-2532

Rammed Earth is a construction company specializing in high-end
residential construction. However, they occasionally offer classes in
rammed-earth construction.

REAL GOODS INSTITUTE FOR SOLAR LIVING
555 Leslie St.
Ukiah, CA 95482-5507
(707) 468-9292

The Real Goods people offer hands-on classes in straw-bail construction
as well as other classes emphasizing energy efficiency, do-it-yourself
residential hydro-systems, solar power, and beer making!

SACRAMENTO OWNER-BUILDER SCHOOL
4777 Sunrise Blvd. Suite A,
Fair Oaks, CA 95628
(916) 961-2453

This school offers instruction on the many aspects of house building.
In a forty-two-hour long session they teach, using slides and demonstra-
tions, everything you need to successfully build your own house.

COLORADO

SOLAR ENERGY INSTITUTE
P.O. Box 715
Carbondale, CO 81623
(970) 963-8850

The Solar Energy Institute is a nonprofit organization that teaches
workshops on solar architecture and technology.

REDROCK COMMUNITY COLLEGE
13300 West Sixth Ave.
Lakewood, CO 80401-5398
(303) 914-6362

This community college offers classroom instruction in all of the
following phases of house building: site preparation, foundation systems,
floor and wall framing, roof coverings, and building code compliance. It
also offers classes in solar construction technology.

FLORIDA

FLORIDA SOLAR ENERGY INSTITUTE
1679 Clearlake Rd.
Cocoa, FL 32922-5702
(407) 638-1000

The Florida Solar Energy Institute offer workshops on passive solar
energy, solar water-heating, and energy-efficient design.

INTERNATIONAL BAU-BIOLOGIE AND ECOLOGY
P.O. Box 387
Clearwater, FL 34615
(813) 461-4371

This nonprofit institute is dedicated to making people aware of both the
physical and chemical health hazards caused by conventional
building methods. Classes teach how to avoid such dangers by using
ecologically-sound building techniques developed over the past two
decades in Europe.

MAINE

FOX MAPLE SCHOOL OF TRADITIONAL BUILDING
P.O. Box 249
Brownfield, ME 04010
(207) 935-3720

The Fox Maple School is well know for its instructors' expertise
in timber framing. It offers hands-on classes in traditional timber
framing as well as instruction in straw-bail building techniques.

SHELTER INSTITUTE
38 Center Street
Bath, ME 04530
(207) 442-7938

The Shelter Institute was founded in 1974 by Pat and Patsy Hannen,
and according to Ms. Hannen, it was the first institute of its kind. In the
past twenty-two years, over 22,000 students have graduated from its
program. With an emphasis on post-and-beam construction, hands-on
classes and classroom instruction are taught for every phase of construction.

MASSACHUSETTS

HEARTWOOD
Johnson Hill Rd.
Washington, MA 01235
(413) 623-6677

Heartwood was established in 1978, and its instructors believe that "the
best way to learn something is to do it." They teach house building in
hands-on, live-in sessions that last three weeks. They also offer courses
on carpentry for women, timber framing, and house design.

Fig. 30. Photo courtesy of the Natural House Building Center in Santa Fe, New Mexico.

MINNESOTA

NATURAL SPACES DOMES
37955 Bridge Rd.
North Branch, MN 55056
(612) 674-5005

Buckminster Fuller would be proud to know there's a school teaching how to put together his geodesic domes—Natural Spaces Domes offers dome construction workshops in Minnesota and Virginia. They also sell dome homes.

GREAT LAKES SCHOOL OF LOG BUILDING
Snowshoe Trail, Sand Lake
Isabella, MN 55607
(218) 365-2126 (612) 822-5955

Great Lakes School teaches everything needed to build a log home, but students are required to provide their own tools. During the course, students can live in a bunkhouse or tent for no added cost. The school brochure asserts that the cost of a good log house can be considerably less than a comparable wood-frame building. The school is proud that it gives people the skills to avoid long-term mortgages.

MONTANA

CENTER FOR RESOURCEFUL BUILDING TECHNOLOGY
P.O. Box 100
Missoula, MT 59806

C.R.B.T. is a nonprofit organization dedicated to teaching about alternative building techniques.

NEW MEXICO

NATURAL HOUSE BUILDING CENTER
2300 W. Alameda
Santa Fe, NM 87501
(505) 471-5314

The Natural House Building Center teaches a variety of basic skills needed to build your own natural home in just one workshop. In addition, they teach the necessity of a healthy "habitat." Workshops are held in Santa Fe, New Mexico as well as in Vancouver, British Columbia; Ashville, North Carolina; Madison, Wisconsin; Canterbury, New Hampshire; Fairfield, Iowa; and Austin, Texas.

SOUTHWEST SOLARADOBE SCHOOL
P.O. Box 153
Bosque, NM 87006
(505) 861-1255

The Southwest Solaradobe School teaches earth building, such as adobe and rammed-earth techniques, at locations across the Southwest. It also publishes the Earthbuilder's Encyclopedia.

BLACK RANGE LODGE
KINGSTON, NM
(505) 895-5652

Black Range Lodge is not actually a school, but it's a great place to visit and checkout a variety of straw-bail structures. It also sells books and videos on straw-bail construction, holistic living, and rammed-earth building techniques.

NEW YORK

EARTHWOOD BUILDING SCHOOL
366 Murtagh Hill Rd.
West Chazy, NY 12992
(518) 493-7744

Earthwood Building School is directed by Rob Roy, who has written a number of books on building houses. Earthwood offers hands-on classes in cordwood masonry, straw-bail construction, and alternative energy.

EASTFIELD VILLAGE
P.O. Box 539
Nassau, NY 12123
(518) 766-2422

Eastfield offers workshops in several traditional crafts, including flatwall plastering, tinsmithing, as well as architectural and ornamental stone cutting.

OREGON

APROVECHO RESEARCH CENTER
80574 Hazelton Rd.
Cottage Grove, OR 97424
(541) 942-8198

Aprovecho Research Center is situated on forty acres in Oregon. Its resident staff, interns, members, and friends teach sustainable forestry, organic gardening, indigenous skills, permaculture, and other aspects of ecological living. Classes are also available in solar-assisted housing (passive and active), earthen structures, straw-bail building, and geodesics.

COBB COTTAGE COMPANY
P.O. Box 123
Cottage Grove, OR 97424
(541) 942-3021

Cobb Cottage Company teaches how to build with cob, a mixture of earth, straw, and sand. Cobb is one of the cheapest construction materials available as well as one of the oldest. In England, there are many cobb homes that have been occupied for hundreds of years. Classes are offered in Oregon, Washington, and California.

GROUNDWORKS
P.O. Box 381
Murphy, OR 97533
(541) 471-3470

Groundworks teaches all sorts of techniques for building earthen homes, which have been used for thousands of years in every possible climate. If constructed with a sturdy foundation and an adequate roof, an earthen home can stand for hundreds of years. In addition, the institute offers classes in straw-bail construction, which blends well with earth-building techniques. Groundworks places an emphasis on women's classes, but, not to worry chaps, coed workshops are also available.

LOST VALLEY EDUCATIONAL CENTER
81868 Lost Valley Lane
Dexter, OR 97431
(541) 937-3351

Lost Valley is an international community and nonprofit educational
center located on eighty-seven acres just outside Eugene, Oregon. The
center focuses on sustainable living skills and offers a broad range of
courses throughout the year on a variety of subjects, such as straw-bail
construction, cobb, organic gardening, permaculture, and community
living skills. Classes range in length from one day workshops to three
month apprenticeships.

PENNSYLVANIA

EAST COAST ALTERNATIVE BUILDING CENTER
2801 Ataxville Rd.
York, PA 1704
(717) 792- 0551

The East Coast Alternative Building Center offers hands-on workshops
for straw-bail construction, mud plaster, stucco, cobb, and other alternative building methods. The center also teaches passive and active solar
design and application.

TEXAS

CENTER FOR MAXIMUM POTENTIAL BUILDING SYSTEMS
8604 FM 969
Austin, TX 78724
(512) 928-4786

This is a nonprofit organization specializing in sustainable design prac-
tices. The center focuses on how to use appropriate materials—i.e.
indigenous or readily available materials—for a variety of buildings and
sites. Members of the Center For Maximum Potential Building Systems
are available for lectures, seminars, and workshops (both in the United
States and overseas).

OWNER-BUILDER CENTER AT THE HOUSTON COMMUNITY COLLEGE
4141 Costa Rica
Houston, TX 77092
(713) 956-1178

Houston Community College offers hands-on classes in foundation
layout, framing, electrical, plumbing, door and window installation,
drywall, tile installation, and roofing.

VERMONT

YESTERMORROW
RR1, Box 97-5
Warren, VT 05674
(802) 496-5545

Yestermorrow provides live-in, hands-on classes in practically every
phase of home building, inside and out—home design/build, wood-
working, furniture design/build, home energy design, landscape design/
build.

VIRGINIA

BEAR MOUNTAIN OUTDOOR SCHOOL
U.S. 250
Hightown, VA 24444
(540) 468-2700

Bear Mountain Outdoor School, located high in the Allegheny Mountains in western Virginia, offers "learning vacations" for just about
everyone. Instruction is given in hand-hewn log building, stone construction, timber framing, sustainable design/build, solar design, and
alternative energy systems.

WASHINGTON

GREENFIRE INSTITUTE
1509 Queen Anne Avenue North; Number 606
Seattle, WA 98227
(206) 284-7470

Greenfire Institute teaches weekend courses in the theory and practice of straw-bail construction. Sessions alternate hands-on building experience with lectures and presentations. Classes are available in both Washington and Oregon.

HABITAT FOR HUMANITY

121 Habitat Street
Americus, GA 31709-3498
(912) 924-6935

"We believe in Habitat's integrity, effectiveness and tremendous vision. With Habitat, we build more than houses. We build families, communities, and hope."
JIMMY CARTER,
Former U.S. President

As a Habitat volunteer, I spent a Saturday digging holes for a much needed structure at a childrens' playground. The next morning I expected to hurt all over, but I didn't hurt at all. So I said to myself "It takes awhile; I'm sure I'll hurt tomorrow." But when tomorrow came I still didn't hurt. I concluded that since I was working for a worthy cause and throughly enjoying every minute there would be no aches or pain.

I wish I had more time to volunteer. It certainly is a wonderful feeling to give your time and muscle to an organization as incredible and gratifying as Habitat for Humanity.

This wonderful ecumenical organization was founded in 1976 by Millard Fuller and his wife Linda. With thanks to incredible volunteer support and donations, Habitat has built over forty thousand houses around the world and has provided decent, safe, affordable shelter to families in need. Habitat invites people from all walks of life to work with the families that will be living in these houses. Perhaps you've seen former President Jimmy Carter hammering nails at one of Habitat's projects.

Volunteer some of your time and you'll get something back. By helping a family in need obtain inexpensive, quality housing, you will get hands on experience, and you'll feel great. Give them a call—there's almost certainly a project in your area that could use your help.

CHRISTMAS IN APRIL

I spent a fulfilling week working with Christmas in April (CinA) and the Castro Street Lions. On the last Saturday of April, 1998, CinA with the help of 4,500 volunteers, fixed up 45 houses for the low-income, elderly and disabled people of San Francisco, as well as 18 schools, community centers, group homes and hospices. I was the captain for one such house belonging to a delightful old lady named Berenice. George, my co-captain, and I went to see what work we could do to make her house a little more comfortable and safe.

We submitted a list of possible improvments to the CinA staff and discussed our project. George ordered painting and cleaning supplies while I picked up lumber, handrails, a door, a smoke alarm and various other items.

Since we were going to paint the back of the house and since the house was quite tall we had scaffolding installed a few days before the weekend. On Saturday, I arrived at the house before 8 A.M. and most of the 15 or so volunteers were already there drinking coffee and eating bagels.

We soon got to work, though, and painted the back of the house, the master bedroom, the spare bedroom, and the hall. Another crew attached handrails to make going down the stairs easier for Berenice. Dave Waldren re-mortered the bricks on the fireplace, which were crumbling, while others cleaned and scrubbed. When we were done, the place looked great and Berenice was thrilled.

Some of us met up later at the CinA picnic. I enjoyed talking to these volunteers. Some of them had done other volunteer work that included building and renovating older dwellings. One or two had built houses in Africa while working for The Peace Corps. I asked them if they thought their volunteer experience could be applied to other areas in their lives. The answer was a resounding "Yes!"

On Sunday, April 26, the sprit of Christmas continued with Sukkot in April. Volunteers from Congregation Sherith Israel and Jewish Family and Childrens Services refurbished 10 homes and three non-profit facilities.

Christmas in April may have a chapter near you. You can phone them at 800 473-4229.

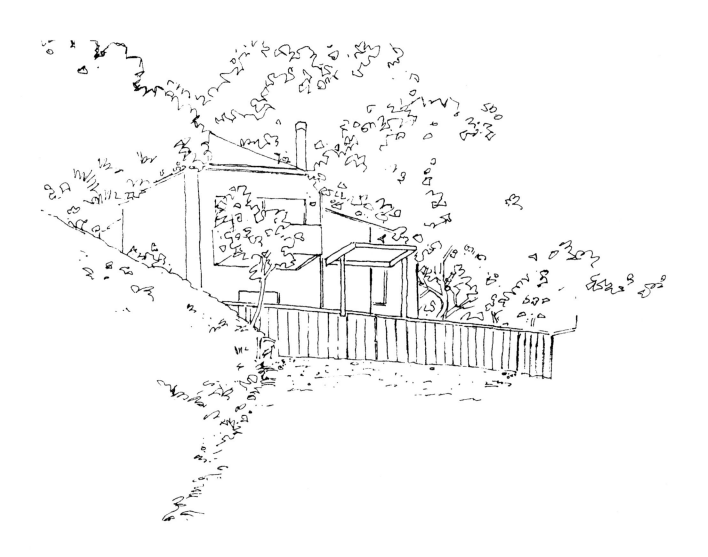

8 HOW WE BUILT OUR DREAM HOUSE FOR A SONG

Though it was some years ago, I remember those first days of building as clearly as anything in my life—after months and months of preparation, we were finally getting to work! The task before us seemed enormous. We were attempting to build a unique and beautiful house, and none of us, except my friend John, had more than amateur experience. Instead of pondering all the 'what-ifs' of failure, we dug our hands in, and day by day, we worked as a team to raise our dream house. I did most of the construction with John, Marcia kept working in our restaurant to keep up the monthly cash flow, and Emily and Liz helped in every little way they could.

BEGINNINGS

It all started with a phone call. Some close friends convinced us to submit an offer on a piece of property adjoining theirs. The land was in a dreamy valley set between golden hills, spotted here and there with oak, laurel, and madrone. The neighborhood was called Sleepy Hollow—pleasant, suburban—the perfect place to raise two daughters.

I remember phoning the real estate agent to enquire about the price and terms. It turned out two other parties were also submitting offers. Marcia and I decided to make an offer $1,000 higher than the asking price. This may seem a little unusual, considering we hadn't even seen the lot, but our friends informed us that there was a possibility the property could be divided into two parcels. The lot was seven acres; all the other surrounding properties were only two or three acres. We were the successful buyers.

Fig. 31. Emily and Liz on the backhoe during the time we were waiting for our building permit to be approved.

We were very glad that our friends had that hunch—which is all it really was—about splitting the property. The first time I enquired at the county planning department about dividing the property, I was told we couldn't split the lot. But I persisted, and eventually, the sale of the newly-created parcel was a very nice bonus.

In that same summer, after we'd bought our land, I'd taken a course for owner builders. After the course, I knew that I could build my own house. In fact, I was so confident of it that I wanted to tackle a project beyond the bounds of conventional building—I was determined to build something unique. So I began a quest. I devoted my spare time to looking for materials at auction sales and used building yards.

I had two criteria for judging materials: cheap and interesting. As a matter of principle, I wanted to spend as little on materials as possible, and so it became a hobby to see how little I could actually spend. For over a year, I collected materials and tools (I even purchased a backhoe). Unusual items prompted me to use my imagination, and soon I was coming up with all sorts of unconventional ways for incorporating all these found materials. Finally, I found the steel building that brought

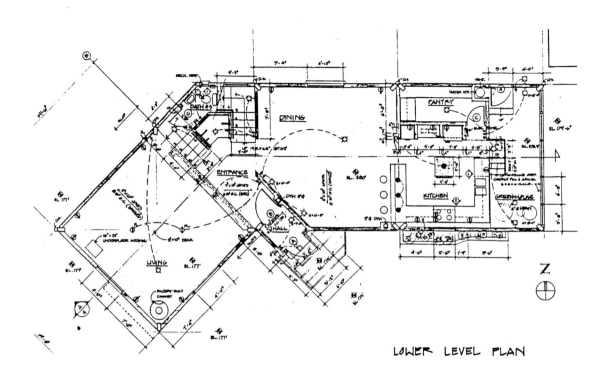

LOWER LEVEL PLAN

Fig. 32. and 33.
Guelimero Marin's
brilliant plans.

the vision of my dream home into focus (the story of which is discussed in detail on page 49-50).

It might seem that all this collecting with only a vague notion of the final outcome was a trifle backwards—perhaps, a severe case of putting the old cart before the horse. But it turned out to be a very excellent order to go about things. By buying materials in an unorthodox and unhurried manner, I was never pressured to compromise my aesthetics because I needed this or that immediately and couldn't afford the best. I considered everything that I was going to use in my dream house, in some way or another, beautiful.

By now, I was certain that I wanted to build with steel—not the easiest among available materials when it comes to design. I knew that I would need the help of an architect. It should be no surprise, however, that I didn't want to pay retail for an architect's help. Instead of going to an architectural firm, I decided to post a notice at our local university. It was a request for a graduate architectural student to design an unusual

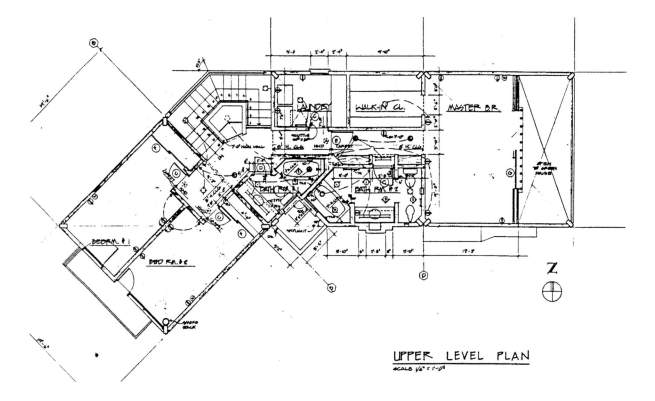

UPPER LEVEL PLAN
SCALE 1/4" = 1'-0"

Fig. 33.

house, the drawings and finished product would make for an excellent thesis. Unfortunately, none of the respondents had the least clue of how to transform my steel building into a house. But an architect from Mexico, who chanced upon my ad while looking for a teaching job at the university, had more than just a clue. Guelimero Marin came up with a brilliant design, using almost all the steel.

In retrospect, I would have to admit that using steel had its disadvantages. It added the expense of hiring an engineer to do certain calculations on the steel frame. Furthermore, erecting the steel was no simple job because we couldn't afford the luxury of a crane. Nevertheless, the steel building provided me with a very unique and highly aesthetic dream house.

After completing the plans, submitting them to the building department, and making the obligatory changes and corrections, we were finally ready to begin. My wife, Marcia, our two daughters—eight-year old Emily and five-year old Elizabeth—and I moved into our newly acquired mobile home.

I was in particularly good spirits in those days, feeling that I had freed us from the financial burden of rent. Since selling our last house and waiting for approval on our building permit, we had been paying almost $1,000 a month. Finally, here we were, living on our own land, rent free, in a comfortable mobile home that I had bought for approximately $3,000.

This is the condensed version of how we built our dream house. By improvising and using our imaginations, we were able to own the home of our dreams, and, hopefully, this story will lend a few ideas to aspiring owner builders who wish to do the same.

Opposite page.
The backhoe
was purchased
at an auction
for $5,000. It
was used
for 5 years and
sold privately
for $5,000.

PREPARING THE BUILDING SITE

Our property was rather hilly, and we wanted a flat area for a garden as well as a place for our children to play. Leveling the building site was a fairly big chore. But considering how much easier it was going to make the building process, we decided that it would be worthwhile.

To create a flat building site and keep the earth where we wanted it, we had to build two retaining walls. This entailed sinking railroad tracks into the side of my hill. Holes had to be drilled to a depth of 11 feet, and all of the holes needed to be 12 inches in diameter. This was one of the jobs that I couldn't do by myself, so I hired a drilling crew to come out and drill the holes.

Then, it was time to lower the steel railroad track into the holes. However, I ran into a problem working alone. The diameter of the holes was so small I couldn't seem to lower the steel into them without disturbing the sides and filling them up. To get through this first impasse, I enlisted the help of my youngest daughter, Liz—who by now had reached the ripe old age of six. We looped a rope around the steel, and she directed each track into its respective hole (fig. 34).

Once all the steel was in the holes, it had to be made plumb, in other words, vertical. When we thought we had the steel standing straight in the holes, we proceeded to fill them with concrete. We checked for plumbness a final time.

Next, we laid railroad ties behind the steel posts to make a wall. We sometimes used nails, driven into the railroad ties and bent around the steel track, in order to keep them in place. We checked each layer of railroad ties to make sure it was level. This was not necessarily structurally important, but it did ensure a good-looking wall.

We placed a drainage pipe behind the wall to carry rainwater away from the house. I used some old salvaged 12-inch and 16-inch corru-

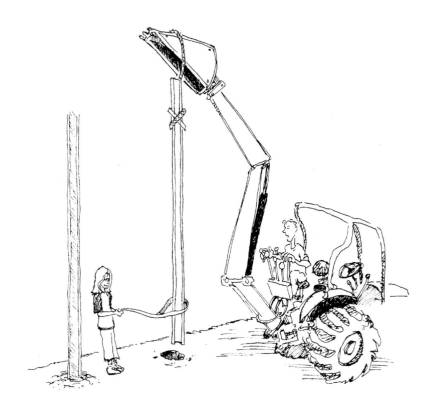

Fig. 34. Liz helped me lower the steel into the holes for the retaining wall.

Fig. 35. Liz and Emily, standing where their new house would be built.

gated steel pipe that I had bought at an auction. I started with the 12-inch and let it run into the 16-inch. This is a good example of adapting your methods to the materials you have on hand. While the engineer I consulted suggested 10-inch pipe, I used larger pipe because that's what I had.

When the two retaining walls were finished, I graded the site again with my backhoe, pushing excess earth down my future driveway. The excess earth was easily levelled out on my 200-foot-long driveway.

FOUNDATION

With the site prepared, we were ready to begin building the house. We laid out its shape on the ground using string and batter boards (fig. 36). Then, we marked the exact location of the piers for the drilling crew. The drilling went relatively smoothly except for one or two holes where the crew hit rock at a fairly shallow depth. The engineer said this was not really a problem since the concrete column would be sitting on solid rock. Because we were building on a hillside where earthquakes are an ever-present threat, our engineer thought that our best and safest choice was a pier and grade beam foundation. I was concerned about the expense, but this choice turned out to be quite economical. As you can see from the foundation plan in figure eleven, we had to drill twenty holes, which really wasn't that much work.

Putting in the foundation was basically a repeat of building the retaining walls. However, in building the foundation, we placed reinforcing steel, called rebar, in each hole, the vertical rebar was then tied together with rebar running horizontally inside plywood forms. After this was finished, we filled this shell of rebar with concrete. We left the center and one end open (see fig. 38), allowing the concrete truck to back right inside the foundation. This eliminated the need for a concrete pump and saved us about $500. The last part of laying the concrete was then completed with the help of my brother-in-law and a wheelbarrow.

The mudsill is the timber joined directly to the foundation. For this final stage of the foundation we used pressure-treated lumber, which is rot-resistant. We leveled the mudsill extra carefully because the rest of our house would be sitting directly on top of it. By the following day, the concrete had set-up. We removed the forms and tightened the anchor bolts that had been imbedded into the concrete to hold the mudsill in place.

Those large masses of concrete that you see in figure thirty-seven were designed to support the steel columns. When the steel beams were first

The rain descended, and the floods came and beat upon the house; and it fell not: for it was founded upon a rock.
MATTHEW V11, 25C. 75

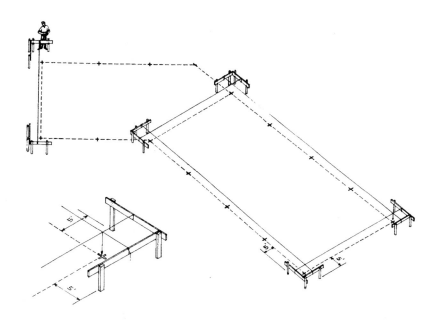

Fig. 36. Laying out the house with batter boards.

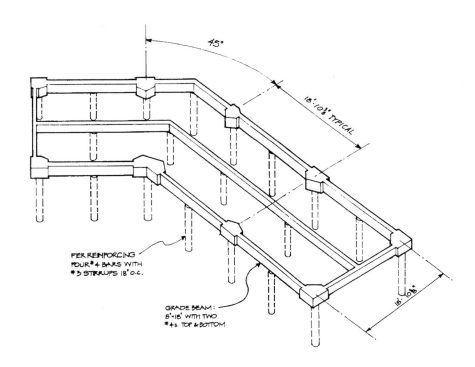

Fig. 37. An Architect's drawing of our pier and grade beam foundation. The large masses of concrete are what the steel columns sit on.

PIER REINFORCING
FOUR #4 BARS WITH
#3 STIRRUPS 18" O.C.

GRADE BEAM:
8"×18" WITH TWO
#4s TOP & BOTTOM

45°

18'-10⅜" TYPICAL

18'-10⅜"

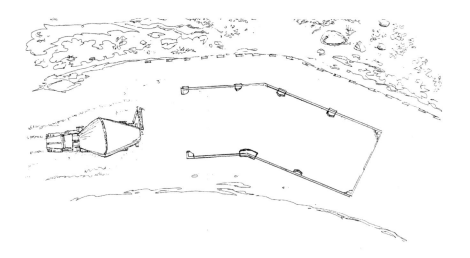

Fig. 38. We left the end of the foundation open to avoid the extra expense of hiring a concrete pump.

Fig. 39. Filling the form with concrete. Notice our retaining wall in the background.

Fig. 40. John and I erecting the first steel.

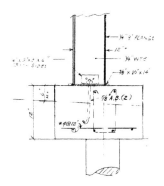

Fig. 40a. Detail of a Steel column support.

erected, the connectors were the only things holding the beams upright (fig. 40a). Once we got the steel connectors in place, we lifted the steel columns into position with the backhoe and bolted them securely.

I acquired the connectors through an unusual exchange at an auction. I purchased a vast lot of large bolts and nuts—many more than I needed, but for such a great price that I just couldn't pass them up.

The owner of a steel fabrication plant, who arrived late at the same auction, approached me and gave me his card. He asked that I keep him in mind should I want to unload any of my bolts. A light bulb went off in my head, and I asked if he would be willing to fabricate the steel connectors that I would soon be needing in exchange for many bolts. Before the end of the auction, we came to an agreement. He sent a couple of his employees to the auction site to pick up the lot of nuts and bolts. This suited me because it meant I didn't have to do any hauling. Soon after, I took a diagram of the steel connectors over to his company. His employees fabricated exactly what I had needed.

It's great when this sort of trade goes smoothly, but remember that trading materials in this way can be a little risky—you never quite know who you're dealing with.

JOISTS

The next job was to put on the joists. Joists are the timbers laid edgewise to provide floor support. They're usually spaced 16 inches on center (O.C.) but sometimes 12 inches and occasionally 24 inches O.C. We started at the left side of the house and marked the position of each joist on the mudsill. Next, we placed the joists on top of our marks and toenailed them, that is drove the nails in at a slant to the mudsill.

UNDERFLOOR PLUMBING

Since the building department required an underfloor plumbing inspection, we assembled all the underfloor drain lines next. We used plastic ABS pipe, which is glued together, and fed it into the 4-inch castiron pipe outside the building. But before we did this we capped the end in order to test the system. We set up a vertical 10-foot pipe as far from the rest of the system as possible. The 10-foot vertical pipe (when filled with water) exerted intense downward pressure, allowing us to test the system for weak connections. I'm proud to say that our ABS joints passed with flying colors.

SUBFLOOR

We used 3/4-inch tongue and groove plywood. To avoid any squeaks, we both glued and nailed it down. Since we were going to use tiles in the entry, kitchen, and greenhouse, we used 11/8-inch plywood that I had bought very cheaply at an auction.

This job went very quickly and cheaply. I'd found a pneumatic-coil nail gun at an auction for $35 and a couple of boxes of full head nails for $20. I recommend buying a nail gun even if it's not at this bargain price. It makes your life so much easier. Honestly, we actually enjoyed putting down the subflooring! Once the subfloor was in place, we had a big level area on which to work.

No man has lived to much purpose unless he has built a house, begotten a son, or written a book.
ITALIAN PROVERB

Fig. 41. John hoisted up a wall using Proctor wall jacks.

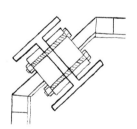

Fig. 41a. How we joined wood members to the steel to achieve a nailing surface.

WALLS

It was now time to put up the walls. In our case, since we were working with steel, our first job was to attach a 2x4 to each side of the steel columns. This was accomplished by using the oxyacetylene torch and cutting or burning holes in the steel. We then bolted a 2x4 to each side, which provided us with a nailing surface (fig. 41a).

We built one wall at a time, starting with the outside walls and working our way in. We assembled the walls on the floor, nailing the vertical (studs) and horizontal (plates) pieces together first. Then we hoisted it up all at once. This is a very exciting part of the building process, because so much happens so fast.

We measured the length of our walls and checked the plans for the height. Then, we deducted 4½ inches for the one bottom and two top plates and cut the studs accordingly. We put the top and bottom plates together and marked the position of each stud, so they would line up exactly. We spread the two plates apart, placed the studs in between, and nailed the whole thing together.

Fig. 42. Emily and Liz couldn't wait to have dinner on the second floor of their new house.

Finally, we hoisted each outside wall and nailed it in place (fig. 18). It didn't take us long to frame the interior walls of the lower level since our design was very open. It's fast going work, and even if you work slowly, you could easily frame a couple of walls a day. The structure was starting to look like a house.

I decided to leave all the plumbing and electrical until the house was completely closed in, thinking it would be best to work on those inside over the winter.

Raising the second floor walls was in principle the same as raising the first—we just had a better view while we worked—so we repeated the process. We put the second floor joists on top of the first floor walls and then nailed and glued down the plywood subfloor.

HOISTING THE STEEL

Before we could complete the wood framing, we had to get the rest of the steel frame erected. To do this we used a come-along, to hold the steel against the loader bucket, and lift it to the second floor. The most

Fig. 43. This steel member was erected single-handedly, thanks to modern lifting devises by the Proctor company, Inc.

Fig. 44. The job of erecting the steel became easier each day.

difficult part was getting the steel members up and attaching them to the vertical columns already in place. John was doubtful that we could do it by ourselves.

I thought about the problem very thoroughly and figured out that by using two Proctor wall jacks I could probably single-handedly erect the first of these steel members. I reasoned that once I had one up, I could use it to help get the next one up and so on. However, the first one was the most difficult going. It took me all weekend, but it was well worth my energy to see John's face that next Monday morning!

As you can see from the photo in figure forty-three, I used two Proctor wall jacks to gradually hoist the beam into place. Once in position, it only had to be bolted down. It was a rather tedious job

Fig. 45. The completed steel frame and all the ground floor framing.

Fig. 46. We erected the second story walls with the siding already attached.

because the beams had to be perfectly positioned. The next beam was a little easier. I attached a block and tackle to the top of the first beam and used it to help me hoist up the second. There were twelve of these steel beams altogether. The most difficult were the two in the greenhouse because there was no floor to work on. I made a temporary platform to work on. I used two pieces of steel channel that were bolted together and then bolted these in turn to the vertical steel columns. Finally, I laid a couple of sheets of plywood on top and secured these to the steel as well.

COMPLETING THE FRAMING

Home wasn't built in a day.
GOODMAN ACE

I accomplished two things when I bolted the steel members in place. I had both put together the main frame of the house and erected the roof frame. Once the main steel frame was put together, it was very easy to assemble the rest of the framework. Now, in my case there were no bearing walls on the upper level, so the roof was self-supporting. In a typical house the walls support the roof, but in my house no headers were necessary, except in the area around the stairs. We finally had something that looked very much like a house—or a masterful steel sculpture.

Before completing the wood framing for the second floor walls, I decided to attach the siding (fig. 46), eliminating the need for scaffolding. Nailing on the siding first actually worked out better than we expected and saved us a bundle.

Before hoisting the second floor walls, we had to make certain that they were going to be square. We did this by measuring the diagonals and making sure they were equal. Then we attached the vapor barrier and nailed on the siding. Using our Proctor wall jacks, we lifted the walls and nudged them into place.

Next, we built the interior walls for the second floor. There were more interior walls upstairs than downstairs, but the method of constructing the walls was the same.

ROOF FRAMING AND RAFTERS

The roof framing was very easy. We used the steel purlins that had come with the steel building instead of rafters. We bolted them every 16 inches, using the oxyacetylene torch to make the holes for the bolts. With the purlins in place, it was time to add the plywood sheathing to the roof frame. Normally, the plywood would have been nailed to the

"It shows what can be done by taking a little trouble," said Eeyore. "Do you see, Pooh? Do you see Piglet? Brains first and then hard work. Look at it! That's the way to build a house," said Eeyore proudly.
A. A. MILNE. *The House at Pooh Corner*

rafters. We bolted the 1/2-inch plywood to the steel purlins. Unfortunately, the shape of our roof caused a great deal of waste, but in this case there was no way to avoid it.

The plywood was the roof deck. We could have used tile or shake to finish the roof, but I used composition shingles that I had bought at an auction for $50. They were red and green. I had been the only bidder on this lot because everyone else had been afraid of the colors. After they were nailed down, I coated them with a roof emulsion so that they looked dark brown. I like to tell my friends that I have the best $50 roof around—not bad, especially on a half-million dollar house!

I'd been hoping to get the framing done before the rains started. In Northern California, it generally rains from late October to early May. A finished roof helps protect the wood members, deterring any warping. We'd managed to finish the framing just in time for spring—so much for timing.

PLUMBING

I found this phase of construction fairly simple. John and I did all the plumbing ourselves. Planning in advance for the plumbing is important. One way to make sure you have enough space for drain, waste, and vent pipes (DWV) is to build bathrooms back to back and use wider framing members. I didn't use any framing other than 2x4s, but we were able to use the space created by the steel. We had to box in one area to accommodate the pipes.

The water lines took very little space, and we used mostly ½-inch pipe for that. But for drain waste and vents, we used 1½-inch, 2-inch, and 3-inch ABS pipe. ABS is cheap, so we had leeway for a few mistakes. By using more vent pipes, I was able to get away with mostly 1½-inch and 2-inch DWV pipe, which was easily accommodated by 2×4 framing. Many builders use a main stack for venting and bring all the vent lines into this one big stack to go up through the roof.

DWV carries away waste from the toilets, showers, dishwasher, and washing machine, so we had to be careful when venting these pipes to allow potentially dangerous gases to escape into the atmosphere. ABS pipe was the only thing we used for our drain lines, because it's a breeze to cut and connect. We cut it with almost anything available—handsaw, *Skillsaw, Sawzall*, chop saw—and glued it to the fittings. If we made mistakes at this stage, it was relatively easy to fix. There were some rules to follow so that it all worked. One very important thing to remember was that once we glued ABS, we couldn't get it apart without cutting it.

WATERLINES

Rather than try to explain how we laid the waterlines, I'll just give you some advice. Practice joining copper pipe or "sweating it together," as it's called. Go to your local home improvement center or hardware store and buy a torch, a length of copper pipe, some fittings, some solder, flux, a small brush, and a pipe cutter. You may use a cheap cutter to practice, but go to a plumbing supply and buy the best before you begin plumbing your house. Also, consider buying a very small cutter for tight corners. You'll also need something to clean the pipe like an emery cloth or some other nifty device.

A few words of caution: do not touch the hot copper pipe or fittings. Use a vise or vise grips to hold the pipe steady until you get the hang of it. And be careful of fire. Quite a few buildings have burned to the ground because of careless plumbers. When working in tight quarters, be extra careful. I always put a tile or a piece of sheet metal behind the pipe to protect the framing and, more flamable still, the paper-backed insulation. Always have water close at hand.

Wet down the area with a spray bottle before sweating a joint. But remember that you have to make sure the inside of the pipe is dry, because if there's any water in the pipe, it won't solder. Working with any pipe that has had water in it can be a real problem. An old plumber's trick is to make a ball with some plain white bread and put it into the pipe to soak up the moisture. It dissolves after you turn the water on.

When putting in the waterlines, you must constantly be thinking about how to maintain the water pressure. Unless you're getting water from a well or a tank, the water comes from a water meter and is fed from that point to the kitchen, bathrooms, laundry room, and outside spigots. Thinking about pressure, we started with a 1-inch line at the entry to the house, reduced it to 3/4-inch line, and finally to 1/2-inch lines towards the end.

FIRE SPRINKLER SYSTEM

Because my house was so far from the street, the building inspectors required that I install a fire sprinkler system. Another surprise expense— I couldn't believe it! I had to hire a fire sprinkler engineer to design the system, but at least the sprinkler heads only cost a few dollars each and I had bought a lot of copper pipe and fittings at an auction, so the actual cost of the sprinkler system was quite low.

More and more municipalities are now requiring homes to be fitted with fire sprinkler systems, especially in areas at high risk to fire. It may sound like an extra job that you don't want to tackle; however, it's much cheaper if you do it yourself. After installing the rest of the plumbing for your house, you'll have plenty of experience for the job.

When we were finishing our house, we found we had a little cash left over, on account of the plumber not knowing about it.
MARK TWAIN

HOT WATER

We used on-demand water heaters. We put one upstairs and one downstairs. The upstairs one is particularly useful for filling our whirlpool tub, which holds about sixty gallons of water. An on-demand water heater is switched on when you turn on the water tap. The flow of water activates a switch so the water is heated as it moves through the heater. With an on-demand heater, we never run out of hot water. They're also more energy efficient since they're only on when needed.

A brand new water heater cost about $400 retail. I actually bought three of these on demand water heaters from a man who was the successful bidder at the auction. He paid $15 each for a lot of over a hundred heaters. I approached him and asked how much he wanted for just three heaters. We finally agreed on $30 a heater. This was a good deal for both of us. He doubled his money, and I didn't have to contend with the hundred heaters left to move. This is another example of procuring goods in an unorthodox manner. Never be bashful about approaching the successful bidder of a large lot, because with a little bargaining, you'll probably strike a deal on just the amount of goods you wanted.

ELECTRICAL

I had no idea of how to put in an electrical system. I hadn't any idea of how to do plumbing either when I started, but at least I'd felt confident that I could do it. This wasn't the case with the electrical work. I hired an electrician. He was out of work at the time so I was able strike a pretty good deal with him.

I'll tell you in a nutshell what I know about electrical work. The service comes into your house either underground or overhead. I elected to bring it in underground. I had my own backhoe and could easily dig the trench. I thought it looked nicer, too. However, it may be cheaper to bring it in overhead.

Once you've brought the service up to your house it goes into the service panel which costs around $100. The electrical company will plug the meter into this panel. Then electricity is distributed to every room in your house via romex, which is quite inexpensive—a 250 foot roll costs about $30. Items like outlets and switches are less than a dollar each. So if you do decide to put in the electrical system yourself, it shouldn't be a major expense.

Fig. 47. This is our free standing fireplace that I bought at the auction for $350; it retails for over $1,000.

Fire is the most tolerable third party.
HENRY DAVID THOREAU.

HEATING AND COOLING

Here again, I knew almost nothing; luckily, John had some experience. I spotted an ad in the local newspaper for a combination gas-forced air heater and a 220 volt air conditioner. John and I went to look at it. As it turned out, the owner had taken two or three units off a commercial building that was to be demolished. Since he only needed one, we obtained a $3,000 unit for $400. I had a load of 8-inch flexible duct, which I had bought at an auction. With all this duct, we were able to run the system to every room.

If you are going to use a forced air heating system, you must think about where your duct work is going to run. Quite often it runs between the joists, and you may have to box in an area inside a closet to go upstairs. We gave absolutely no thought to this problem until it was time to do it, and in our case, it ended up being okay because we had a natural space created by the steel in which to run the ducts.

Fig. 48. The hardwood flooring pattern that our architect suggested, "start at the middle of the room with a one inch square and work out."

We were required to specify the size of the heating unit on our plans. I just drew in a rectangular box under the house and specified the minimum size required by the company that did the energy calculations. Though everything worked out well, I do suggest doing a little more planning than we did.

FIREPLACE

Our wood burning stove came from an auction at a waterbed store that went out of business. I paid $350 for it, and it could almost heat the entire house on its own. We tiled the area under the wood burning fireplace with tiles I'd bought at an auction.

FLOORS

I had been worrying about what we were going to use for flooring. I couldn't believe it when Marcia showed me the ad. Here it was in the newspaper: 1-inch thick tongue and groove maple flooring for 75 cents a square foot. I phoned immediately. I thought it was a mistake, but it wasn't. The man down at Caldwell's building wreckers said it had been salvaged from the old Rincon Annex Post Office in San Francisco.

The maple was very rough, but the guy at Caldwell's used building supplies showed me some that had been sanded. It was beautiful. In retrospect, I should have run it through a surface planer before putting it down, but at the time I didn't even know what a surface planer was. We just cleaned up the tongue and groove with a wire brush and nailed it down using my finish nail gun.

Actually, it was a little more complicated than that. I wasn't sure which way to run it, so I made the mistake of asking my architect. Figure forty-eight shows the pattern he recommended. When I look at it, I think of the patterned floors in the Louvre. We used this same pattern in the living room and dining room and then switched to a traditional lines-going-straight-across pattern on the second floor. We managed to cover all the floors that needed it with this handsome flooring.

The balance of the floors were done in high quality tile from Brazil. I paid $50 for more than 600 square feet of Brazilian tile at an auction at a bankrupt tile company in Santa Cruz. We had used 11/8-inch plywood in the dining room, kitchen, and entry. Since I'd bought a lot of glue from the tile auction, we decided to glue it all down.

We also glued tile down in the greenhouse, it only had a subfloor of 3/4-inch plywood. But if you use tile, make sure you have a very solid surface to lay it on. In our case, it seems to have been fine because the joists were every 12 inches, O.C. Otherwise, I advise using a thin set mortar, which is a cement base rather than a glue.

INSULATION

Since our house is basically a wood frame house with high ceilings, good insulation was essential. The living room has a 12-foot high ceiling; the dining room and kitchen have 9-foot high ceilings; and the three bedrooms have 15-foot high ceilings. The house is a bit expensive to heat, but we love all the space. Of course, we're lucky to live in a warm climate.

Fiberglass insulation is very easy to work with. You just cut it to fit between the studs, rafters, and joists. To cut the fiberglass in straight lines, lay a piece of wood on top of it. Although there are many kinds of fiberglass, paper-backed fiberglass is the most commonly used. The paper acts as a vapor barrier, and there's a fold of paper along the edge that you staple to the wood members. It's really very simple, but make sure you wear a respirator and long-sleeved shirt. The fiberglass particles get into the pores of your skin and are very itchy. I wouldn't want to think about what the fiberglass particles do to your lungs without a respirator.

DRYWALL

Back in my naive-building days, I used to think that sheet rock was required on all walls and ceilings because of potential fire hazards from unprotected wood surfaces. The reality is that sheet rock is just a very inexpensive product for covering walls and ceilings. It's used in place of plaster, which requires more skill. Sheet rock is so inexpensive that it's the one thing that even I don't recommend buying at auction, unless you're ready to use it right then and happen to pick up a lot cheap. I dislike sheet rocking; in fact, it's on the top of my "I don't do" list. One day, I'm going to build a house without a single square inch of it.

But, back to the house at hand—when the time for sheet rock arrived, I placed an ad in the local newspaper for a drywaller and received a number of replies. One man was particularly eager to do the work. He wanted to come right out to see the job.

Joe arrived in an orange *International* van, accompanied by his two dogs. He explained that he was a drywall specialist from Washington state and desperately needed work. I hired him on the spot. He was very grateful to get the job. I had four acres, and my brother-in-law had a large tent he wanted to get rid of for $50. I let Joe and his dogs live on my property, and in turn, he insisted on working for an hourly rate that was $2 less than what I'd originally offered him. He also was adamant about using drywall screws instead of nails "because it made a much better job." I have to say that it looked great when he was finished.

Fig. 49. This was our original idea for the east wall of the greenhouse with 5 lower windows.

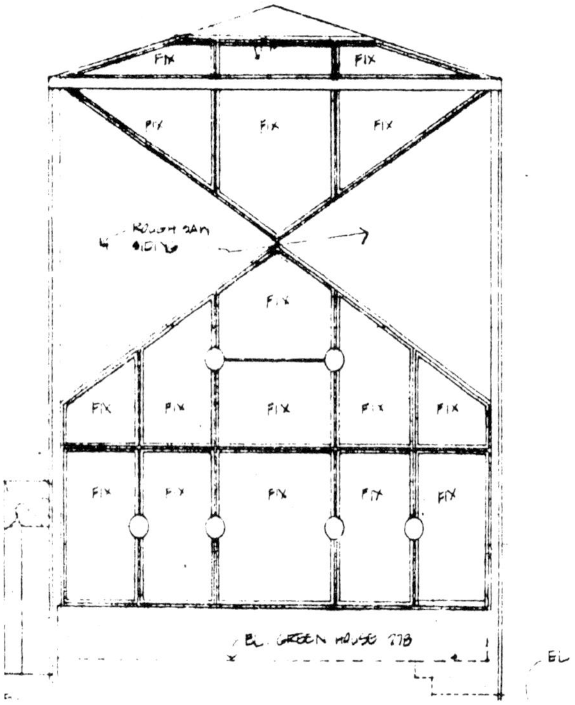

Fig. 50. We accommodated 7 of the $10 windows. The wall was made in one piece on the upper floor, and lowered into position.

Fig. 51. Our kitchen cabinets were imported from Germany. You can see the Brazilian tile bought for $50.

WINDOWS

During the construction of our house, I made a great buy on a quantity of double-glazed, bronze-aluminum windows for which I paid $10 each. The sizes were approximately the same as those specified on the plans, so it made sense to use them. When I bought them we hadn't yet framed the walls, and it was an easy matter to accommodate them. Windows for new construction like the ones we used have a flange, making them very easy to nail into the rough opening. In the greenhouse room, we made a major modification to the east wall. Here again a little ingenuity was in order (fig. 49 and 50).

DOORS

We used commercial doors, another great buy—$10 each from an auction. They had metal skins, but that didn't pose a problem as they were predrilled for hinges. John and I went down to the hardware store and bought hinges. I had bought a supply of door framing material from the same auction we bought the roofing material, so it was economical to

*Fig. 52. The finished
master bathroom.*

make our own frames. It was pretty easy, too, since John had some
experience. However, pre-hung doors, already mounted in their frames,
are available at most home improvement centers. And if you can't find
doors you like at auction, pre-hung ones aren't such a bad idea.

PAINT

Use a good quality paint. We made the mistake of buying some inferior
paint and had to go over many areas twice. We used a cheap airless
sprayer, which performed quite well, but next time I would rent a more
expensive sprayer.

We did the painting after the drywall, windows, doors, and the
finished floor had all been installed. We didn't have to cover the maple

floors because they still needed a great deal of sanding. However, we did have to cover the windows and door frames. The doors were spray-painted outside.

CABINETS

The kitchen cabinets were purchased at a furniture show in San Francisco. My father-in-law is in the furniture business, and he informed us that a German company, exhibiting at the show, wanted to sell their displays cheap rather than bear the cost of shipping them back to Germany. That was several years before we had even bought our property. The cabinets had remained stored in my father-in-law's warehouse until it was time to use them. I tell you this only because I want to emphasize the advantage of having access to free dry storage!

We installed the lower cabinets first then the counter top, that had come with the cabinets as part of the display. We installed the upper cabinets in one area, but didn't use all of them because we wanted a unique look. Instead of cabinets, we had decided to put in a big picture window. We still have plenty of storage because we'd put in a pantry.

The cabinets in the master bathroom were handmade by John. I'd picked up a load of maple wood from an auction—much wider than the flooring. I had so much of the maple that I even used it to surround the Jacuzzi tub. And when I was done with the bathroom I still had enough wood to build our fence (which was kind of a shame but at least I didn't waste it). I only paid $150 for the whole lot!

TRIM

The trim came from an auction that took place at a home improvement center closing down to make way for a new city hall. There was enough window casing to do all the windows and the baseboards. We adapted window casing material to use as baseboard by using a smaller piece along the bottom edge to hide the gap. It all looks quite professional, especially considering what we paid for it. We did break our "don't pay retail" rule when we bought some more elaborate baseboard for the living and dining rooms, we'd saved so much money that we were feeling a little extravagant that day.

MASTER BATHROOM

Since we were anxious to move in, we left the master bathroom empty. The building department made a separate building permit for it, and several weeks went by until we finished it. This wasn't really inconvenient because we had another full bathroom as well as a half bathroom already finished.

APPLIANCES

We paid retail for our convection oven and six-burner cooktop. We used an old refrigerator for a while, then one Sunday I spied an auction ad in the newspaper for a bankrupt appliance company. I went over and picked up a new refrigerator for half of retail. They also had convection ovens and cooktops, but not the ones we wanted. I feel that the finish items in your dream house are not to be compromised. You should get what you want, after all, you deserve the best—even if you have to wait a bit.

FINISHING

We replaced the formica counter tops in the kitchen with black ceramic tile over a year after we moved in. Also we changed many of the original lighting fixtures. This is easily done but quite expensive. We just waited until we had the money to buy what we wanted. Other than that, everything that originally went into our house was quite satisfactory.

MOVING IN

We removed our mobile home on the day we had our final inspection. However, before the planning department would give the okay to the power company to turn on our electricity and gas, I got a call from the planner. Apparently, we had forgotten to do some landscaping that we'd promised to do.

This was two days before Thanksgiving, and we had already invited everyone we knew to come to the new house for dinner. I explained this to the planner, but she would not make any exceptions. I offered to post a bond of $2,000, insuring that I would finish the landscaping in thirty days or forfeit the bond. This convinced the planner. She promised to phone the utility company that afternoon.

A storm rolled in that night, and our temporary power pole was blown down after two years of faithful service. The day before Thanksgiving, I waited anxiously for the power company to arrive. Morning passed by and I was still waiting. Around noon I phoned them, and they assured me that they would make it over. The electricity division arrived at about three o'clock, installed the meter, and turned on the power. It was wonderful to see the house come alive for the first time—like some sort of brilliant, fortuitous sign. Then, I waited and waited for the gas people to arrive so that we could put the gas cooktop, water heaters, and furnace to work. At around five o'clock they finally turned up and hooked up the gas. It had all worked out fine.

The happy ending: we threw a great Thanksgiving party and had much to be thankful for on our first day in our dream house!

GUEST HOUSE AND CAR PORTS

I went on to build a guest house, consisting of a studio with a full bathroom and a small deck. I finished the three carports and extensive landscaping. John decided to retire for good this time. I had learned so much from him that I was able to do most of the work myself. Time was no longer a factor, so these projects were done on a part-time basis over the next couple of years.

CONCLUSION

There are some things that I would do differently next time. For one thing, if I wanted another big house, I would build it in stages. I would do more of the work myself. And finally, I would try to waste less building materials. All said and done, it was a wonderful experience and I have a great sense of security and independence knowing that I can always build another house anywhere anytime.

Fig. 53. The finished house.

*After falling in love
and witnessing the
birth of your children,
building your own house
is the most gratifying
thing you can do.*
DAVID COOK

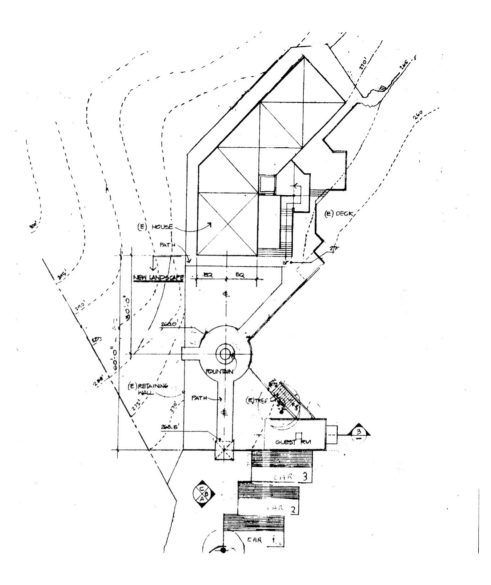

Fig. 54. This is the landscape plan, showing the main house, the guest house, and the three carports.

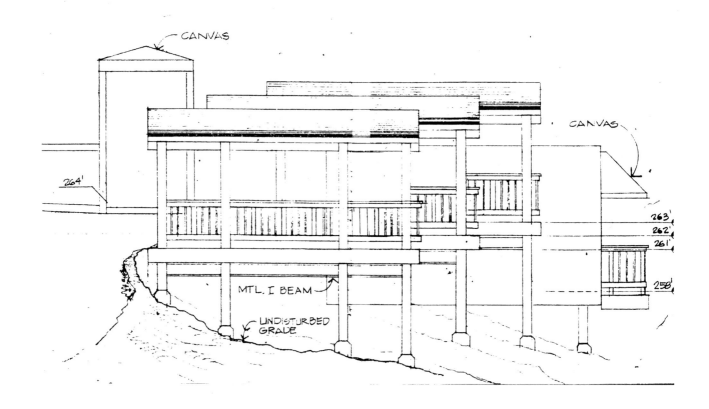

CANVAS

CANVAS

264'

263'
262'
261'

258'

MTL. I BEAM

UNDISTURBED
GRADE

GLOSSARY

Assessment	A government fee, added into the price of a piece of property to cover the expense of a road, sewer, or other improvement.
Adz	A tool similar to an ax with the blade set at a right angle to the handle.
Batter boards	Horizontal boards used to fasten strings indicating outlines of a foundation.
Block and tackle	A rope and pulley often used to redistribute weight in order to lift materials or tools, (e.g. used to lift steel beams into place or a bathtub to a second story).
Cap plate	Uppermost plate running perpendicular to the studs.
Come along	Rachet tool with chain (or cable) and hook used for pulling.

Easement	Right of way over someone else's property.
Grade Beam	A horizontal, reinforced concrete beam, resting on the ground between two supporting piers in a foundation.
Ingress and egress	The right to enter and exit.
Joist	A framing timber laid edgewise as floor support.
Lien	A claim on a piece of property as security of payment of a debt.
Lot	1.) A quantity of materials sold together as one unit at an auction 2.) A parcel of land.
Mortise and tenon	A joint created by a mortise, an incision cut to snugly receive the tongue-like tenon sticking off the end of a timber, plank or board.
Mudsill	A timber attached directly to the concrete of the foundation; the house-frame attaches directly to the mudsill.
On center (O.C.)	A term used to determine spacing from the center of one stud, rafter, or joist to the next.
Oxyacetylene torch	A torch fueled by a mixture of oxygen and acetylene gas that creates an extremely hot flame for all kinds of metalworking.
Plat map	A map of a piece of land that has been divided into building lots.
Pneumatic tool	A tool powered by compressed air, e.g. nail gun or paint sprayer.
Purlin	A horizontal timber (or piece of steel) used to support the roof.

Redemption period	The time during which back taxes may be paid on a tax certificate to clear any lien for which the property may be auctioned off.
Romex	Generically, used by electrical workers to refer to any nonmetallic sheathed cable; actually, the trade name for type NM cable produced by General Cable.
R values	The insulation factor used to determine heating efficiency of different building materials—the higher the R value of a material, the more resistant it is to letting energy pass through.
Shill	Someone hired that pretends to bid and buy to drive up the prices at an auction.
Stud	Vertical timber used in a series to support walls; generally, nailed to horizontal plates.
Surface planer	A machine for producing flat, smooth boards or surfaces of wood.
Tongue and groove	A joint made from a board with a protrusion or tongue that fits snugly into the groove of an adjoining board—often used to join hardwood flooring or plywood.
Tract-built	Describes houses that were built all at once in a planned subdivision.
Wattle and daub	Mud or clay trawled onto a work of sticks interwoven with smaller twigs and branches;

SUGGESTED READINGS

Ching, Francis D.K. *Building Construction Illustrated.* New York: Van Norstrand Reinhold, 1991

Curren, June Norris. *Drafting House Plans*, Sacramento, CA: Brooks Publishing Co. 1991

Kardon, Redwood. *Code Check.* Tauton Press 1995

Roskind, Robert. *Building Your Own House,* Berkeley, CA: Ten SpeedPress. 1984

Sher, Les and Carol. *Finding and buying your Place in the Contry.* Chicago, Ill: Dearborn Financial, 1992.

Starr, Ronald. *Buying Real Estate Super-Bargains at California Tax Sales.* Unlimited Golden Opportunities Press, P.O box 27218 Oakland, CA 94602

Wagner, Willis H. *Modern Carpentry.* South Holland, Ill: Goodheart-Willcox. 1983

Fine Homebuilding. (Magazine)

Journal of Light Construction. (Magazine)

BIBLIOGRAPHY

Tunis, Edwin. *Colonial Living*: New York. Thomas Y. Crowell Company 1957

Dennis, Marshall. *Residential Mortgage Lending*: W. Reston Publishing Company, Inc 1985

Broderbund Software Inc, *3D Home Architect*, 500 Redwood Blvd, Novato, CA 94948-6121 (415) 382-4600

The Detail Book E.C.P Publications 478 West Hamilton Ave. Suite 151 Campbell, CA 95008. Tel 1 800 823-5503

Leibowitz, Sandra. *Eco Building Schools*: 2765 Potter Street, Eugene, OR, 97405 email: sleibowitz@aaa.uoregon.edu

Sardinsky, Robert. and the Rocky Mountain Institute. Resource-Efficient Housing: *An Annotated Bibliography and Directory of Helpful Organizations*. Snowmass, CO: The Rocky Mountain Institute, 1991 edition.

Shelter Institute, 97
sheet rock, 131
siding, 123
site preparation, 110
site plan, 76
sod housing, 11
Solar Energy Institute, 96
Southwest Solar-Adobe School, 100
stairs, 66
steel: building with, 111, 119-124; buying, 50;
designing   for, 110; hoisting steel members, 120-123
steel connectors, 117
storage: dry, 46, 133; outside, 46; free, 46, 135
subfloor, 118
straw-bale housing, 14, 94, 95
sweating pipe, 125

tax-defaulted land sales, 23-29
tax certificates, 23
tax certificate states, 31
tax deeds, 23, 30
tax deed states, 31
temporary dwellings, 10, 80-87
thirty-year mortgages, 36-38
tile: Brazilian tile, 49; laying, 130; tile glue, 130
timber-frame house, 10, 11, 97
tongue and groove flooring: buying from a wrecking
   yard, 53; economical floor covering, 69; patterns, 130
tools: buying at auction, 52; for scavenging (list), 55;
   electrical, 52; pneumatic, 52;
tract-built housing, 13, 93
trusses, 64

underfloor plumbing, 118

vapor barrier, 123
vicinity map, 76

walls, 119
water heaters, 127
water lines: advise for sweating pipe, 125
mobile home, 85
windows: installing, 133; trim, 135
wrecking yards, building, 53

Yestermorrow, 103

zoning, 19, 109